The Ocean World of Jacques Cousteau

Window In the Sea

The Ocean World of Jacques Cousteau

Volume 4

Window In the Sea

THE DANBURY PRESS

This human tourist slowly gliding his way through the gorgeous pastures of this shallow reef is liberated from his own weight, enraptured by effortless mastery of a three-dimensional world. But here he has not quite penetrated the true ocean. Around him sunlight still casts shadows, vivid colors endow the scenery with life. If he were to dive—in a bathyscaphe—to 2000 feet or beyond, he would be surrounded by eternal darkness. With the exceptions of the fitful self-illuminations of some abyssal animals and bacteria, the light man sees by in the deep sea is the light he himself has carried into the depths.

The Danbury Press
A Division of Grolier Enterprises Inc.

Publisher: Robert B. Clarke

Production Supervision: William Frampton

Published by Harry N. Abrams, Inc.

Published exclusively in Canada by
Prentice-Hall of Canada, Ltd.

Revised edition—1975

Project Director: Peter V. Ritner

Managing Editor: Steven Schepp
Assistant Managing Editor: Ruth Dugan
Senior Editors: Donald Dreves
                    Richard Vahan
Assistant Editors: Jill Fairchild
                    Sherry Knox

Creative Director and Designer: Milton Charles

Assistant to Creative Director: Gail Ash
Illustrations Editor: Howard Koslow

Production Manager: Bernard Kass

Science Consultant: Richard C. Murphy

Printed in the United States of America

234567899876

LIBRARY OF CONGRESS CATALOGING
    IN PUBLICATION DATA
Cousteau, Jacques Yves.
    Window in the sea.

    (His The ocean world of Jacques Cousteau;
v. 4)
    1. Underwater light.   2. Underwater
exploration.   3. Marine biology.
I. Title.
[GC181.C68 1975]   620′.416′2   74-23995
ISBN 0-8109-0578-7

Contents

Perhaps the most radical respect in which the world of the sea differs from the world of air concerns man's most vital sense: his vision. For natural light not only soon loses intensity in the ocean but changes in many of its qualities. THE LESSONS OF LIGHT must be learned if we are to understand the marine world and move about within it in comfort.

WHY IS THE SEA BLUE? (Chapter I). To answer this deceptively easy question, we must examine the physical nature of light—and what happens as it strikes and descends into the ocean. For light is affected drastically by the watery medium. It is bent, absorbed, refracted, and reflected, and its colors are subtracted one by one as it passes from the surface to the aphotic zone.

We learn more from looking at objects than from any other form of investigation. This is MAN'S DIRECT VISION (Chapter II). But looking is not as simple a process underwater as on land. Our eyes are complex organs that are designed to receive and interpret light traveling through air. Underwater the human eye does not work so satisfactorily. But with various kinds of aids, from the unsophisticated face mask to the mobile diving saucer, man can see probably better than fish ever did. Recently, too, man has gained the ability to make sustained observations in the sea by going down in habitats to become a member of a reef community, accepted as such by his aquatic neighbors.

Mechanically supporting man's direct vision is only a beginning. There is also MAN'S INDIRECT VISION (Chapter III): photography, television, cinematography. Since 1893, when the first successful underwater photograph was snapped, immense forward steps have been taken in solving the difficult problems of designing lenses, housings, multi-purpose vehicles, lighting, and miscellaneous equipment that can withstand the pressures and corrosive chemistry of seawater and permit man to record in permanent form the life of the sea.

Light, of course, performs more functions than serving as the vector of man's vision. It is THE BEAM OF LIFE (Chapter IV), the visible expression of the sun's energy that is captured by the process of photosynthesis in green plants, which occurs almost exclusively in the upper 100 feet or so of the sea. It is the spark of all organic existence.

Light, the "beam of life," daily coming and going as the earth revolves about its axis, imposes certain rhythms on the sea, a PULSE OF THE OCEAN (Chapter V). One of the most impressive of these is the vertical migration of the deep scattering layer, or DSL, clouds of marine life composed of hundreds of species which rise toward the surface at night and descend 1000 feet or more during the day.

However, all light in the depths is not that artificially produced by man and his gadgets. There is also LIGHT ALIVE (Chapter VI), the bioluminescence of the natural inhabitants of these dark regions. Perhaps three-fourths of all deep-sea species can produce their own light.

Man has been the PROMETHEUS (Chapter VII) of the abyssal depths. With flash powders and flashguns, stop-action and strobe-lighting equipment, laser and holograph apparatus, the contemporary underwater explorer and photographer brings his own sun into regions whose colors will never be touched, or revealed, by the natural sun.

Our interest in light finds a focus in a survey of EYES (Chapter VIII) under the surface, organs that have evolved in almost as many directions as have the creatures of the sea themselves. Some animals in the sea are blind; others possess proportionately huge eyes that probably sense only the presence or absence of light; others scan the entire volume of the water sphere surrounding them: forward and backward, up and down, left and right in a 360° arc—*simultaneously!*

Given these conditions, there is a survival premium placed on FOOLING THE EYE (Chapter IX). Countershading and other forms of camouflage are common; quick-change artists— those animals whose pigmentation automatically takes on the patterns of their backgrounds—abound. However, many of the animals of the sea are color- and light-manipulated.

But, when all is said and done, what is meant by THE BRIGHT PALETTE (Chapter X) that man finds in the darkest, coldest recesses of the sea? What is the purpose of all this color, this beauty—which are never never seen, because the light to reveal them never reaches them? We ponder the problem, and we appreciate the glorious vistas our artificial illuminations open up to us, but we come to no conclusion.

As we return to the surface we reenter the kingdom of light, THE GIFT OF THE SUN. As the ocean is the mother of all life, so is the sun the parent of that energy.

Introduction: The Lessons of Light

Early before sunrise on a clear winter morning, when the horizon reddens above a dark purple Mediterranean, I can see from my home in Monte Carlo the clearly delineated mountains of Corsica, more than 100 miles away. And at night I can see stars that are thousands of light-years away from me.

But when I dive in the clearest of seawaters, the greatest distance at which I can identify an object is 100 feet. With or without any kind of artificial light a diver is imprisoned, as far as vision is concerned, in a bubble of perception only a few dozen feet in diameter. Most fish, most marine animals, are at best in the same situation, but they have developed other senses to extend tremendously their "spheres of perception," and those senses are mainly based on acoustics, pressure waves, or smell.

Underwater man has good hearing, but no echolocation system, no nerves sensitive to slight pressure waves, no smell at all. When diving, man suddenly realizes and appreciates how overwhelmingly dependent his race is upon its sense of vision. He realizes why nothing else—not the loss of hearing, not the loss of an arm or a leg—can be compared with the loss of sight. Throughout this book we will discover how unmatched the sense of vision is (whatever its limitation in water) and how absolutely indispensable light from the sun has been—at the origin of life and still—in producing the quantity and variety of living beings with which we share this planet.

Binocular vision gives man detailed information about shapes, colors, and distances. When I look at a robin singing, I know it is a small bird with a red breast sitting at some 70 feet from me. A blind cat would only know a bird was singing in such-and-such direction. A bat would know a lot more about the shape and the distance of the bird thanks to its echolocation system, but would have no information about its color. Orcas have, like bats, an extraordinarily accurate echolocation device, enabling them to identify their prey hundreds of feet away at least. But when a victim enters their sphere of vision, hunters make good use of the precious additional information given them by their eyes.

Years of direct or indirect visual exploration of the seas with aqualung, with photographic sleds, with diving saucers or bathyscaphes, or with Edgerton's automatic cameras have revealed to me that the ocean is like a gigantic layer cake with occasional raisins or cherries included in it. The overall horizontal stratification of the sea comprises layers of water, transparent or turbid, that are either of a different temperature and salinity, or in which dead particles remain in suspension, or in which planktonic creatures struggle for their minuscule lives. The fruits of the cake are concentrations of larger living creatures, like schools of squid or clouds of shrimp, each one a rare oasis in a three-dimensional liquid desert. But the concentrations disperse and regroup farther away, and the layers constantly move up and down, under the influence of the sun. Everywhere in the open ocean, at dawn, trillions of tons of creatures sink from the surface to the "twilight zone" hundreds of meters below, as if shy of light. After sunset all the multitude rushes back to the surface. This huge daily vertical migration, the "pulse of the oceans," is triggered by light alone—light, the mother of life through photosynthesis; light, the architect of beauty.

Madeira is a Portuguese island 350 miles off the coast of Morocco in the Atlantic Ocean. Its 400,000 inhabitants depend, for their food, upon a "deep-sea monster," the espada (*Aphanopus carbo*). The espada is a ten-pound abyssal fish, the shape of a barracuda, with a bronze-black color and enormous reddish eyes. It is caught by local fishermen at night only, on hooks baited with squid, at 3300 feet if the sky is overcast, at 5000 feet by starlight, but as deep as 7000 feet if the moon shines. These abyssal fish behave according to almost insignificant changes in nocturnal light levels at the surface. How can the difference between starlight and moonlight be felt at 7000 feet of depth?

We know that light entering the sea is absorbed and scattered. Below 100 feet all is deep blue, shadows no longer exist. Below 2000 feet human eyes are unable to detect any light. Yet thousands of feet under the surface, in a world of eternal blackness, colors are revealed in the spotlights of a bathyscaphe. The animals of the abyss still possess pigmentation, mainly gorgeous reds and purples that could never have been seen by any fish since creation? Why? Probably pigments just happen to look red or yellow while accomplishing some other nonvisual function. The sense of beauty they convey to the brain of the human beholder is purely gratuitous. The universe far transcends what man can sense, what he can organize into thought, what he can assign a purpose to. Man receives into his brain only a few narrow bands of the gigantic spectrum of messages dispensed by the cosmos. Outside these narrow bands, only slightly widened by technology, man senses nothing and understands very little.

Jacques-Yves Cousteau

Chapter I. Why Is the Sea Blue?

Why is the sea blue or green? Hold up a glass of seawater and it appears colorless. But stand on the deck of a ship and the sea will appear to have color. Why? To understand it, let us briefly outline what water does to light.

Visible white light is made up of a spectrum of all the colors—red, orange, yellow, green, blue, indigo, violet. When we look at an object and see it as blue, we are seeing the

> **"In many coastal areas, at widespread intervals, red blooms of a species of dinoflagellate become so numerous they cause the water to appear red. This is the famous 'red tide'."**

blue light of the spectrum reflected from the object. All other colors are absorbed and cannot be seen. In the case of the sea, red light is absorbed as soon as it breaks through the water's surface. And by a depth of about 25 feet virtually all the red light discernible to the human eye is gone; a bright red air tank on a diver, for example, would seem a dull dark brown. At a depth of 75 feet a yellow air tank looks more greenish blue, because the discernible yellow light has been absorbed by the water. The still shorter rays of light are almost all absorbed by 100 feet. All that remains are the shortest rays: blue, indigo, and violet. Below 100 feet or so, all light appears a monochromatic blue. So, when the sea is pure and clear, as often is the case in the open ocean, the least absorbed shade of the spectrum, blue, is also reflected to our eyes.

The sea isn't always blue. Some seas appear bluish green or green or brown or even red. These colorations are partly due to the reflection of clouds, but are caused mainly by various particles, mineral or organic, that are in suspension in the water. In some areas, especially along coastlines and in shallower seas, organic matter decomposes and produces a yellow pigment which when mixed with blue light makes the sea appear bluish-green or green. Brown coloration may be caused by sediments, stirred up from the bottom, hanging in suspension in the water and reflecting its brownish coloration. In many coastal areas, at widespread intervals, blooms of a species of dinoflagellate that is red in color become so numerous they cause the waters to appear red. This is the famous "red tide."

When light passes from air to water—that is, from a relatively thin medium to another medium 800 times denser—its speed is reduced from about 186,000 to 140,000 miles per second. Crossing the surface, for the same basic reason, light is bent. This is called refraction. Each color of the spectrum has a different wavelength; blue has the longest and bends the most. The shortest, red and violet, are less refracted. Refraction accounts for magnification when we look through an aquarium glass plate (see page 27).

Penetrating the sea, light is not only refracted and absorbed. It is also scattered, slightly by the water molecules (some scattering occurs even in distilled water) but mainly by particles in suspension.

The sea off Monterey, California, is often clear, blue—and beautiful. But the blue color it reflects is a sign that the waters support only a little life. For it is the more transparent water that reflects the blue and absorbs the other colors of the spectrum. In more nutrient- as well as pollution-filled waters other colors emerge, overshadowing the rich deep blue.

The Penetrating Blue

On a beautiful, sunny day in the tropics, you and your diving companion gather your masks, fins, and snorkels and head for the multicolored reefs. You swim from a beach of white sand to offshore colonies of coral. And there you take a breath and surface-dive down, down, 10, 20, 30 feet below the surface. You stop and look back up at your

diving partner and all is blue. Even her brilliantly colored bathing suit and skin tones have become permeated with blue. Here and there a glint of silver may penetrate down to your level. But mostly the colors are shades of blue—because as the brilliant sunlight passes from air to water above you the warm colors are absorbed and mysterious cold shades take over the undersea world. You are in a new experience!

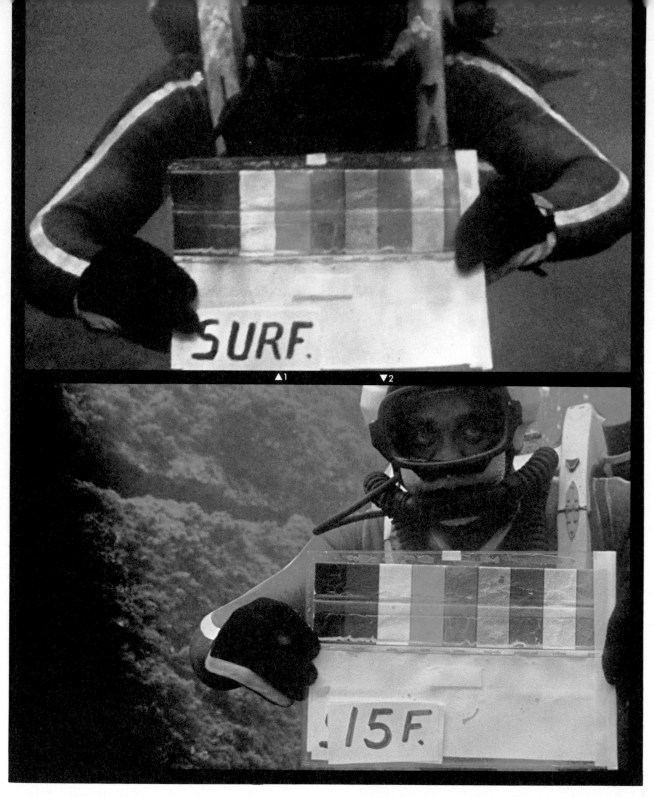

When Colors Fade

The effect of the filtration of light by water is shown in this series of photographs taken of a color chart. Just under the surface the colors appear as in sunlight. At 15 feet the red has begun to fade. At 30 feet the red seems closer to a brown and the yellow is fading. Green and blue predominate. At 60 feet the red has disappeared. The yellow is greener (compare with the first photo), even in the striping on the diver's suit and on

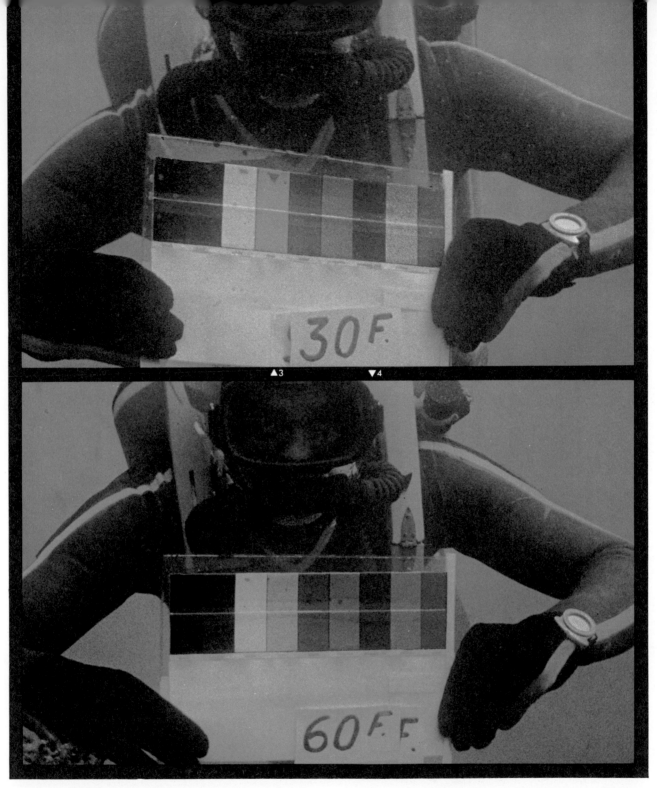

▲3 ▼4

his tank. The green is beginning to take on a bluish hue as compared with the first photo. And the blues are more intense—other colors aren't there to screen them. The white section, from left to right, also shows how the blue comes to predominate as the diver carries the chart to greater depths. This is because white is a combination of all visible light and as some of the visible light is screened out by the water the white segment of the chart reflects the remaining color. The black section does not change.

In Natural Light

In both clear and turbid seas the colors of objects begin to change immediately as we descend. The changes are gradual, and we are sometimes not aware of them. The change and eventual loss of colors are caused by the selective absorption of the many wavelengths of sunlight as it passes through water. Here a photograph has been taken using only the natural light of the sun filtering through water. It appears as though it had been taken through a piece of blue-green cellophane.

16

With Artificial Light

Here is the same scene photographed with artificial light. The light from the flash is white, like sunlight, and contains all the colors of the visible spectrum. The small distance the light traverses from the source to the subject and then back to the camera is short enough that the red, yellow, and others colors are not absorbed by the water. Objects appear in their true colors, or in the colors they would have if they were out of water. With light, the scene takes on a new, crystalline vitality.

With a Flash of Light

The red colors of the squirrelfish, the yellows that tint the soft coral, the brilliant green of the alga show clearly in this underwater photograph because the photographer used artificial light. Without the flash the picture would be a monochromatic blue or blue-green; only the blue sponge in the upper right next to the alga would have retained its true color. If the photographer had taken the picture from 20 or more feet away from

the squirrelfish, it would have appeared a dull brownish color. In the clear waters between Yucatan and the offshore island of Cozumel, visibility is frequently more than 80 feet. Regardless, warm colors are soon absorbed.

"Without the flash the picture would be a monochromatic blue or blue-green. Only the blue sponge would have retained its color."

Turbid Waters

Sediments stirred from the ocean floor along Mexico's Pacific coast cloud the normally clear waters in this aerial view of a coastline.

Plankton, the microscopic plants and animals of the sea, sometimes grow in such profusion they cloud the waters, as in the underwater picture of a California abalone fisherman (opposite). Turbidity caused by

plankton is beneficial to animals. They filter it from the sea directly or depend on it to feed the animals they prey on. These same filter feeders are fouled by the inorganic particles of pollution-caused turbidity.

"Along the more heavily populated coasts of the world, industrial and urban pollution make waters murky."

Crystal fog. At left, fog, scattering the sun's rays, makes it difficult to see details in the landscape. As the day progresses, however, the sun will burn off the fog.

Salt crystals. Above is a painting of what happens when light hits one of the many facets of a salt crystal. Part of the light is reflected onto another facet, that light is reflected onto yet another, etc.

Scattering of Light

Tiny droplets of water suspended in air scatter the light of the sun, obscuring the landscape. In air this crystal fog will soon be dispersed by the sun's rays. Particles of sand, salts, and minerals similarly suspended in water scatter light rays in the same way, bouncing them from one to another until their energy is spent.

Underwater, nothing can disperse the fog. It varies in intensity, but it is pervasive. In harbors or in shallow waters or in waters that have been churned by a storm, turbidity is at a maximum and visibility at a minimum. Even in the middle of the ocean visibility of 300 feet is exceptional.

The scattering process changes the distribution of light, limits the distance one can see, and reduces and subdues direct penetration of sunlight. In water 100 feet deep there are no shadows.

The ocean varies in turbidity from place to place, but one constant characteristic of seawater is that it is salty everywhere. Whether the salinity is high (as in the Red Sea which is surrounded by hot, arid lands) or low (as in frigid polar seas), there are always some particles of salt suspended in the water. The effect upon light of these minute grains is seen in the painting above. When a light ray hits a facet of a salt grain, several things happen to it. Part of the light is reflected and will travel on to another salt particle, whence it will be reflected once more.

Chapter II. Man's Direct Vision

Dive into the sea, open your naked eyes, look around. What do you see? Fuzzy shapes, completely out of focus. No detail. Washed-out colors, hardly anything recognizable. Out of the water, light passes through the air and into your eye—an eye containing a fluid similar in density to seawater. The difference in the density of air and this fluid bends or refracts light rays as they enter your eye. The refracted light focuses on the back part of your eye, the retina, which senses light and passes images onto the brain. Underwater, however, light passes from seawater into your eye bending very slightly because of the similar density of the two fluids. The light rays are not refracted sufficiently to focus on the retina. So everything appears out of focus. You become extremely farsighted.

If you interpose a pocket of air between water and your eye, there is refraction when light passes from the air pocket into your fluid-filled eye. Your eye lens functions properly. Result: you see clearly underwater. To form this air pocket, you don a diver's face mask. With this air lens formed by the face mask, you see almost as well as you do on land. Nevertheless, a mask is not a perfect instrument. The refraction of light through a flat surface of separation between water and air, like the plate glass of a mask or of an aquarium, has the effect of magnifying everything we see by 33 percent. A fish seen 12 feet away appears to be only eight feet away and looks the size you'd expect it to be if it were at that eight-foot distance. Additional problems in underwater vision include tunnel vision and distortion of peripheral vision.

Then, refraction through the mask (plane diopter) increases with the angle of incidence; thus you have undistorted vision in only a narrow beam perpendicular to the plate glass. Lateral vision is heavily distorted, creating a fuzzy image. Unconsciously a diver turns his head constantly from side to side, even more than a hammerhead

> "Even a face mask is not a perfect instrument. A fish seen 12 feet away appears to be only eight feet away and looks the size you'd expect it to be were it at that eight-foot distance."

shark does as it swims along; he keeps his face plate directly in line with the direction in which he is looking.

A mask not only reduces the field of sharp vision, but also drastically shrinks the field of peripheral vision, from more than 180° in air to less than 80° underwater.

What if you need glasses on land? You can wear contact lenses inside your face mask or you can have your prescription ground onto one side of a pair of glasses. The other side, remaining flat, can be glued to the inside of the face plate.

These inconveniences could be corrected by replacing the flat window by a spherical lens. This is successfully achieved with the correcting lenses of underwater cameras. Unfortunately such spherical masks give a different picture in each of your eyes.

The face mask is truly the passport to the sea. Without it man's eyes are unable to focus. It puts an all-important layer of air between eyes and water. In addition, it protects the eyes from irritating salts and toxic organisms contained in seawater.

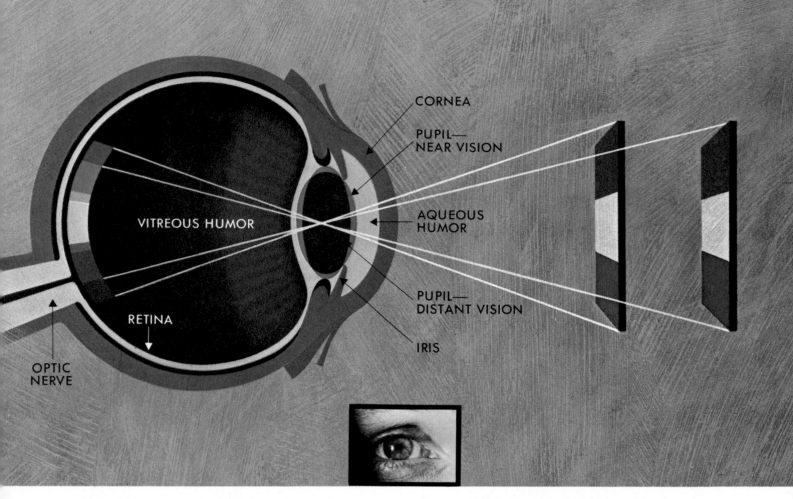

The illustration is labeled with: CORNEA, PUPIL—NEAR VISION, VITREOUS HUMOR, AQUEOUS HUMOR, PUPIL—DISTANT VISION, IRIS, RETINA, OPTIC NERVE

Anatomy of a Human Eye

The human eye is a ball of living tissue about an inch in diameter—a remarkable optical instrument. Among other things it is capable of sensing colors, lights, and shadows; it can tell the form and size of objects; it can judge movements and distances. When light rays strike an object, some of them are absorbed and others are reflected into the atmosphere. If the reflected rays hit the eye, they pass through its clear front window, the cornea. From there they move through the round pupil, the lens, and the clear jellylike tissue behind the lens known as the vitreous humor, and they finally focus on the retina.

At the back of the retina are photosensitive cells that convert the image into a message for the brain. Some of these are visual cells of two types: rods, which function as detectors and enable us to see at night, and cones, which are specialized to catch colors. When we change our gaze from a distant to a nearby object, the lens changes the focus.

The convex spherical cornea does not change its curvature. Being of a watery consistency, it bends the light rays from the direction of their path through air and brings them into

> "Among other things the human eye is capable of sensing colors, lights, and shadows. It can tell the form and size of objects. It can judge movements and distances."

focus on the eye's sensitive retina. With naked eyes a man cannot see clearly underwater because the similarity of water and eye tissue does not allow light rays to be bent sufficiently to focus on the retina.

Water Magnifies

Water is 800 times denser than air. This greater density gives it a different optical quality. The most obvious difference near the surface of the sea is the magnification of objects viewed underwater. The mother and daughter, both partly immersed, look bigger underwater than in the air. Similarly, things divers see underwater appear to be one-third closer than they really are. We must consider this factor in judging size, distance, and mass of anything beneath the surface. Cameras too "see" everything 33 percent closer unless special corrective lenses are used. Divers can, theoretically, be fitted with masks or contact lenses that compensate for water's magnification.

The Difference a Mask Makes

If you were to put your face in the water and look at this sea star without a face mask or goggles, it would appear blurred, as you see on the left. To get the same effect in the air, you would have to look through a powerful $+18$ or $+20$ diopter lens. If, instead, you donned a face mask and then looked at the same sea star, it would snap into focus.

Helmets and Masks

Man must place an air lens before his eyes to see clearly underwater. And he has placed a variety of them there. The helmet with its glass ports was an early method of providing underwatermen with in-focus vision. But it is big, cumbersome, potentially hazardous. And if the face plate should fog with moisture while he is underwater, the man inside cannot wipe it clean.

The five divers at the right demonstrate the versatility of design in face masks. Two wear goggle-type masks that are especially well-fitted for insertion of the corrective lenses of those who wear glasses topside. One diver wears a wide mask which admits more light, although images at its periphery are distorted. The other two divers wear more conventional face masks —one with provision for pinching her nose to equalize pressures. Fogging of face masks can be prevented with silicone spray. If they should become fogged underwater, it's a simple matter to flood and rinse.

Early Devices for Underwater Observation

In seeking to inspect what lies beneath the sea's surface, man has devised all manner of inventions.

One of the early designs, not the earliest by many centuries, was a bell by Franz Kessler. His 1616 design is shown at left. A diver stood in a harness inside a bell which came down to his ankles. His face was at the level of the ports near the top. Air, trapped inside, was what sustained him. What appears to be the bell's clapper at the bottom was a ballast weight.

The second illustration from the left is of John Ernest Williamson's photosphere. From it, with its 30-foot-long, three-foot-diameter flexible metal tube, Williamson in 1914 filmed the first screen version of the Jules Verne classic *20,000 Leagues Under the Sea*. Williamson sat with his camera in the chamber at the bottom and shot through

the port the artificially illuminated action taking place outside. Williamson developed the photosphere from his father's invention of a flexible metal tube system for use in submarine engineering.

Simon Lake's *Argonaut Junior,* second from right, was a wooden submarine on wheels. Lake, a pioneer American submarine builder of the 1890s and early 1900s, built the 14-foot-long, five-foot-high submarine in 1894 and successfully dived her several times. He propelled the craft, which had slight negative buoyancy for rolling over the level bottom, with a hand crank. He waterproofed the two layers of yellow pine planking by sandwiching waterproof canvas between. He built an airlock at the bottom so the crew could step out wearing bucket helmets or just dabble their feet in the water. Compressed air in the boat operated the lock. He adapted the compressed air tank from an old soda fountain. After touring New York

Bay submerged in 1895, he was able to parlay that success into a company that built the *Argonaut,* a 36-foot iron version of *Junior.* The larger *Argonaut* incorporated the first successful snorkel in a submarine.

The first real look into the aphotic or lightless part of the oceans was taken by Dr. William Beebe, director of tropical research for the New York Zoological Society. Beebe won wide acclaim for his record-breaking descents with the bathysphere's designer, a New England engineer named Otis Barton. In the lightless depths of the ocean off Bermuda, Beebe saw and studied fish and other deep-sea creatures never before observed alive in their natural habitat. Barton's bathysphere carried with it several pieces of sophisticated gear of a sort used even today aboard untethered submarine vehicles. And it included lights. Telephone lines as well as a holding cable ran from bathysphere to surface.

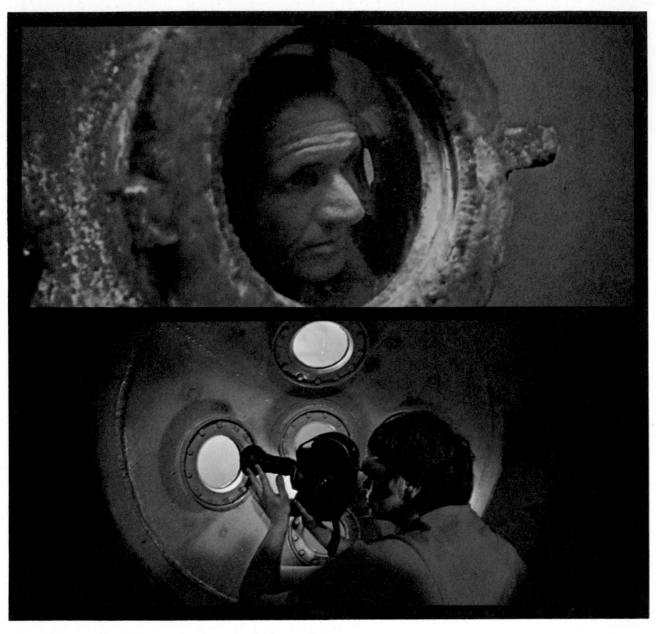

Calypso's Observation Chamber

With the installation of a unique underwater observation chamber in front of the bow of *Calypso* we were able to seek shipwrecks, observe and photograph sea life, or navigate through treacherous shoal waters. The special steel chamber was bolted onto the wooden hull of our 140-foot, YMS-class minesweeper during a complete refit of the vessel at the shipyard in Antibes, France, after we had acquired her in Malta. Access to the mattress-lined chamber is through a 30-inch-diameter entry tube that runs vertically from forepeak to the chamber, eight feet below waterline.

Of the five circular ports for underwater viewing, two face forward, one looks down at a 45° angle and one each faces toward port and starboard. The new bulbous bow added half a knot speed to *Calypso;* it resembles the Maierform bow of modern vessels.

Diving Saucer

Although the aqualung had enabled us to explore the ocean down to the "twilight zone," there were limitations on the length and depth of our dives. In pondering these problems we realized the need for a unique and specialized undersea vehicle. The result was the diving saucer, or *soucoupe plongante* —a radical departure from the traditional submarine and one that would enable us to explore deeper and longer than with the aqualung. We could still be almost as maneuverable as a diver. The six-foot-diameter, five-foot-high circular vehicle has ten eyes—three optical ports for overhead

viewing, two forward-looking ports for pilot and passenger, two photo ports, and three sonar ports—beamed up, down, and forward. Jet nozzles powered by special batteries provide propulsion and lateral steering in any direction, while mercury ballast helps steer the saucer upward or downward. The pilot lies comfortably on a mattress with all

Pioneer vehicle. The easy-to-maneuver diving saucer (above) has six viewing ports and three cameras that we can focus wherever we travel.

controls at fingertips, including buttons for cameras, lights, and tape recorder. To see, to explore, the pilot can "turn the saucer" almost as easily as a diver can turn his head.

37

Manned Undersea Station

Divers outside and oceanáuts inside Conshelf II's Starfish House work in close cooperation. The real question is to demonstrate that men with hands and eyes cannot be replaced by remote controlled instruments. Here Professor Vaissiere and oceanauts from *Calypso* observe at length life on a Red Sea reef. The undersea village off the Sudan

coast also included a garage for the diving saucer, which was thus the first submarine to operate from an underwater base. A "deep cabin" at the 90-foot mark enabled divers to make prolonged observations in depths down to 330 feet without concern for time-consuming decompression. Since Conshelf II, many weeks-long projects involving undersea farming have achieved reality. The potential is unlimited.

Edalhab. *From this underwater station, large enough to live and work in, Dr. Morgan Wells and his wife, among others, carry on experiments.*

Experimenting. *At right Dr. Wells compares a protected piece of coral with unprotected coral to test for the damage done by pollutants.*

Scientific View

Dr. Morgan Wells of the University of North Carolina, working with his wife, lived in the undersea dwelling called Edalhab located in the Florida Keys.

Dr. Wells is shown measuring respiration and photosynthesis by algae growing on this protected piece of brain coral to compare it with data he accumulated from a portion of reef exposed to industrial effluents and pollution. By measuring rate of growth and respiration and comparing figures, he could ascertain the effect of pollutants on the reef. Other scientists also use Edalhab for research requiring repeated observations.

The research carried out is part of project FLARE, Florida Aquanaut Research Expedition, a major project under the jurisdiction of the National Oceanic and Atmospheric Administration in the United States Department of Commerce. Edalhab was first used in the cold waters of New Hampshire's Isles of Shoals by University of New Hampshire marine biologists.

Living in Greater Depths

Ten million square miles of the ocean floor are covered by 600 feet of water or less. The shallower of those depths have been invaded by inquiring goggled men, but depths beyond 300 feet are not so readily within reach. To explore and exploit the sea's riches at those greater depths we devised Continental Shelf Station III, a checkered sphere-shaped dwelling placed in 328 feet of water off France's Mediterranean coast. From Conshelf III's shelter six divers worked, ate, slept, and were monitored electronically from the surface. They showed man could work efficiently even in the hostile cold dim-

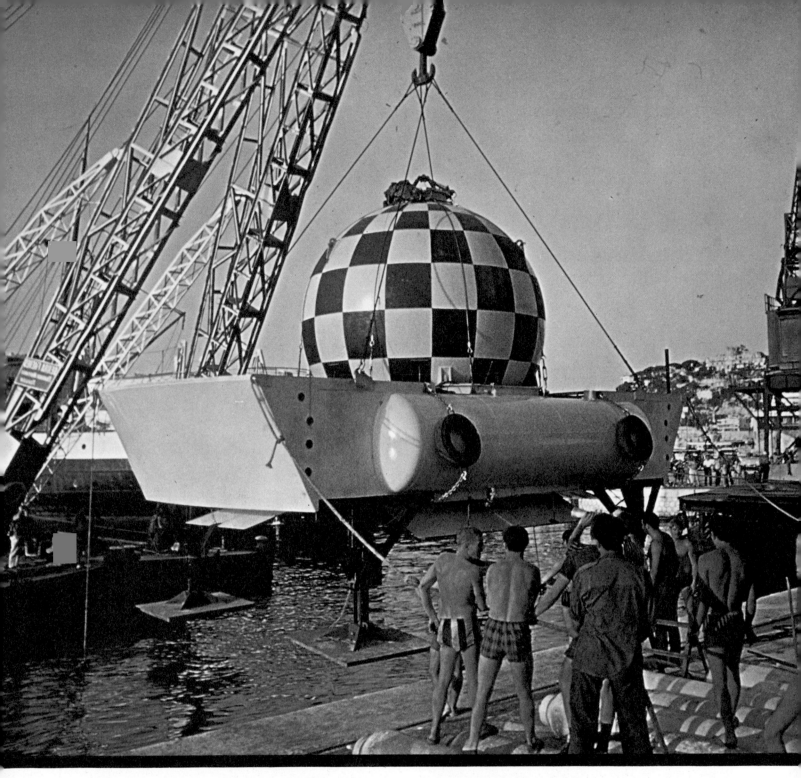

ness of the depths. For our divers took with them the priceless tool—their sight.

Even in the dark waters beyond 300 feet, and with the aid of artificial light, vision was the necessary factor in showing man could, for example, set up an oil-well head. When they weren't outside working, "oceanauts" lived inside their checkered house. At work, they ranged outward some 200 feet and down to depths of 370 feet. Slightly more than three weeks later they surfaced, having proved man could not only exist but could do useful work at such depths—opening up new worlds to his enterprise.

Into the Abyss

Taking man's eyes into the deepest known part of the ocean was the ultimate triumph of the United States Navy's bathyscaphe *Trieste*. This undersea dirigible was a two-man pressurized spherical chamber sus-

pended from a large buoyancy chamber containing nearly incompressible aviation gasoline. Earlier versions of Auguste Piccard's bathyscaphe had been built by the Belgian National Scientific Research Fund (*FNRS-2*) and the French Navy (*FNRS-3*). In 1960, *Trieste* carried the designer's

son, Jacques Piccard, and Lieutenant Don Walsh, USN, 35,800 feet deep into the Mariana Trench of the Challenger Deep, 200 miles southwest of Guam. The 14-mile round trip took nearly nine hours. The men landed on a bottom of fine sediment which engulfed them in a white cloud. There they

***Bathyscaphe* Trieste.** *The photograph above was taken moments before the Trieste's historic seven-mile-down dive. The ship in the background was the Trieste's destroyer escort.*

found sea life—a flat fish resembling a sole and a small shrimp. It was the equivalent of finding life on the moon.

Chapter III. Man's Indirect Vision

Almost 60 years after Louis Boutan took the first photograph underwater, the inventor of stroboscopic lamps and stop-action strobe photography, Harold Edgerton of MIT, applied his art to marine situations.

> "After 1943 the aqualung gave divers freedom and mobility and allowed them to make true motion pictures underwater."

He joined *Calypso* during the summer of 1953 with new equipment adapted to withstand abyssal pressures. Flash and camera units were fastened to a steel frame that finally evolved into a deep-sea motion picture camera sled called the troika that could be towed along the floor of the sea at great depths.

Since the early years still and motion-picture photography have come a long way both above and below the sea's surface. After 1943 the aqualung gave divers freedom and mobility and allowed them time to make true motion pictures underwater.

> "In 1956 our first full-length underwater commercial color film, *The Silent World*, introduced audiences to the startling beauty of the sea's inhabitants and landscapes."

With fast film and strong artificial lights it became possible to make them in color. In 1956 our first full-length underwater color film released commercially, *The Silent World*, introduced movie audiences to the startling beauty of the sea's inhabitants and landscapes. Today underwater photography is the hobby of thousands and an invaluable tool for marine scientists studying underwater life, geological formations, and other subaquatic phenomena.

One of the major steps toward this new and broadening interest was the invention in 1962 of the *Calypso* camera by Belgian engineer Jean deWouters d'Oplinter, with some of the photographers of the *Calypso* team assisting. This is a still camera requiring no cumbersome watertight housing and consequently much easier to handle than earlier equipment. Photographic lenses have improved as well; new ones have been applied to underwater filming.

The use of television cameras under the sea has become widespread. The United States Navy watched the results of the nuclear bomb tests at Bikini Atoll in 1947 on closed-circuit television screens. Television aided in the search and location in 1951 of the sunken British submarine *Affray*.

First successful underwater photograph. This picture of a spider crab amid algae in the Mediterranean waters off Banyuls, France, was taken by Louis Boutan in 1893. To make the photograph, Boutan went underwater with his invention, a fixed-focus camera in a watertight copper box. Eight metal screw clamps closed the lid against the box and a heavy rubber gasket sealed it. There were three openings on the front of the box, also sealed. One was for the lens, the others for the viewfinder. In the rear was provision for changing photographic plates—roll film hadn't been invented yet—while underwater. On the side was a rod for releasing the shutter. The whole device was very heavy. Boutan secured the case and camera on a tripod. Not having full confidence in the watertight integrity of his underwater housing, the professor pressurized the case by attaching an air-filled balloon to a tube running into the housing. The air was forced into the housing by water pressure as he descended.

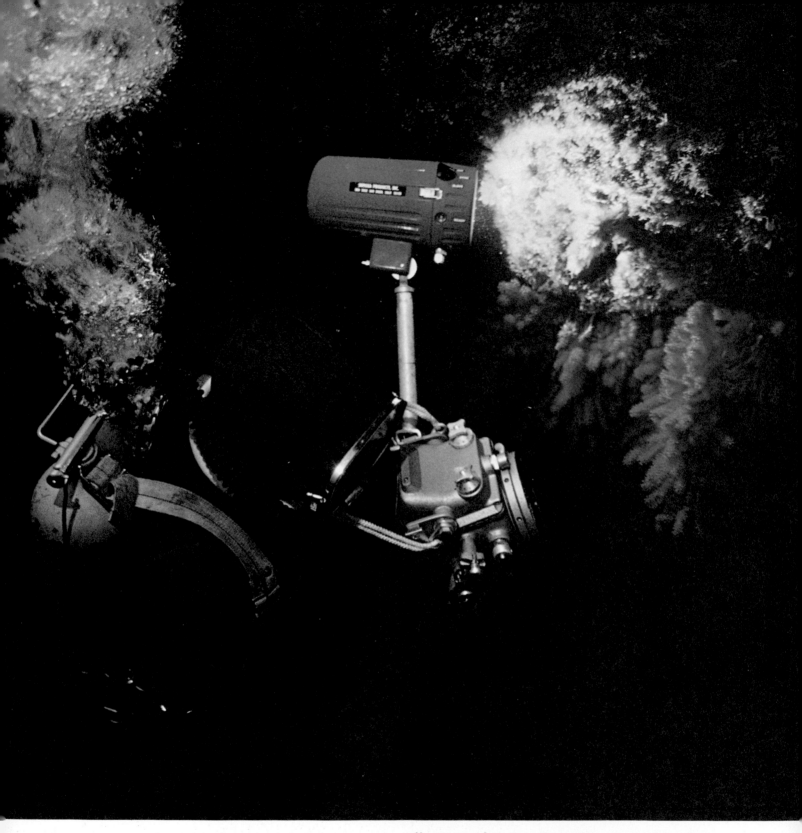

Underwater Still Photos

Recording what the eye sees on film, this diver will have a permanent memory of what he is observing. More and more sport divers prefer shooting pictures rather than spearing fish. Undersea cameras have become standard equipment for scientists or commercial divers. One diver protects his camera from the sea with a plastic housing

he has built himself. Other housings for cameras are made of cast metals treated to resist corrosion. Controls for the camera through the watertight housing are sealed with O-rings—as is the main opening of the underwater case. Some photographers use the special *Calypso*-type camera developed by Belgian engineer Jean deWouters d'Oplinter. That camera requires no underwater housing.

Underwater Television

Very often a human presence will cause fish to act in abnormal ways. We cannot be sure that what we observe is valid. Closed-circuit television allows us to observe life in the sea without intruding upon it. Here a slender television camera is lashed to a tripod near a fish trap and transmits pictures to a monitor in the underwater habitat Edalhab. Marine biologists of FLARE, Florida Aquanauts Research Expedition, kept the fish trap under constant surveillance without influencing the function of the trap or the behavior

of the fish near it. Underwater television has found other applications as well. It is used to make inspections of dams, bridges, and other structures. When we were excavating the ancient Greek wreck at Grand Congloué Island near Marseilles, we used a television unit which gave the venerable archaeologist Fernand Benoit an opportunity to participate in the dig. To cut through the turbid water and make the image sharper on our television screen, we put a large cone of distilled water directly in front of the camera's lens. The results were entirely satisfactory, our reception much sharper.

▲A

▲B

Lenses and Accessories

The optical effects of various camera lenses can be used for various photographic views and studies. In this series of pictures of a Pacific blue sea star the photographer has used different lenses or accessories. The picture of the entire sea star, for example, was shot from a two-foot distance with a wide angle 18-mm lens.

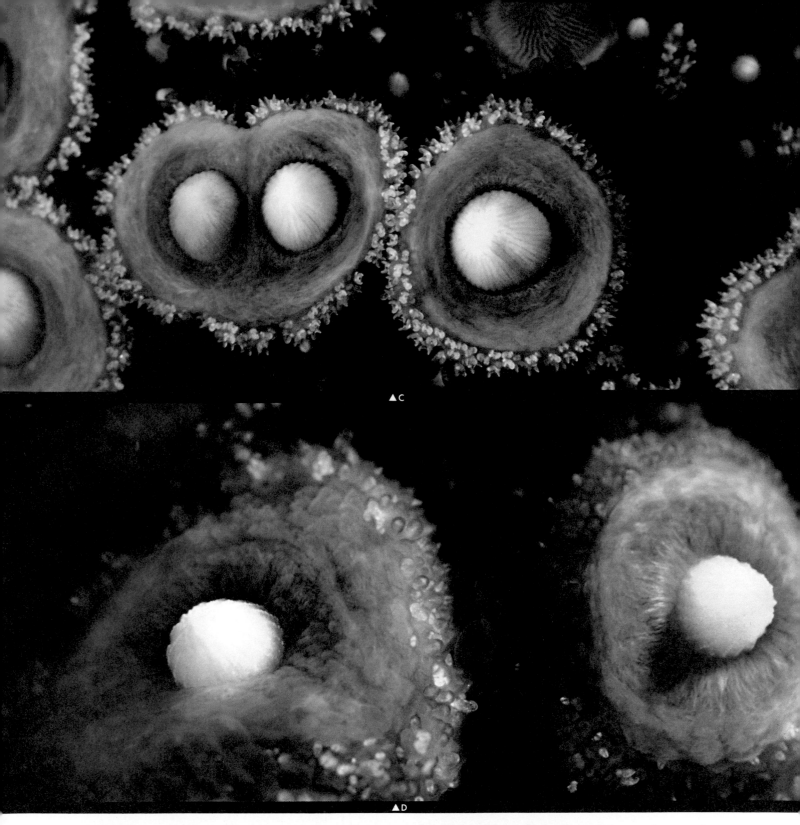

▲C

▲D

A / Wide-angle lens. *This lens allows us to see the entire sea star.*

B / Normal lens. *Shot from only six inches away with a 45-mm lens, this photograph displays details of the sea star's arm we missed with the wide-angle lens.*

C / Extension tubes. *Accessories such as extension tubes to the normal, 45-mm lens show us the arm in even greater detail.*

D / "Closest" close-up. *Using a 32-mm microscope objective, the photographer is able to take a finely detailed close-up excellent for study.*

53

Deep-Sea Camera

On our way to the United States to participate in the First International Oceanographic Congress in 1959, *Calypso* towed this deep-sea camera sled, the troika, across part of the Atlantic; it was equipped with two synchronized still cameras and a strobe designed and built by Harold Edgerton. We were able to bring to the meetings hundreds of color stereo pairs of pictures of the rift valley 7000 feet deep in the center of the

Mid-Atlantic Ridge (a mountain range rising from the ocean floor, emerging here and there to form islands like the Azores). The troika was designed to right itself if tipped over by obstacles. It was the first ever designed for on-the-move filming of the abyss.

We also used a troika equipped with a 16-mm movie camera holding 1000 feet of film and capable of working in 16,000 feet of depth. Right, a sponge-covered slope of an Atlantic seamount—one of the innumerable features of a landscape few ever see.

Mid-Atlantic Ridge

Using our mechanical "eyes," a movie camera and strobe light unit mounted on the troika deep-sea camera sled, we were able to present the photograph above to the First International Congress on Oceanography in New York in 1959. It was a revelation. It showed pillows of lava newly extruded from the center of earth, perhaps

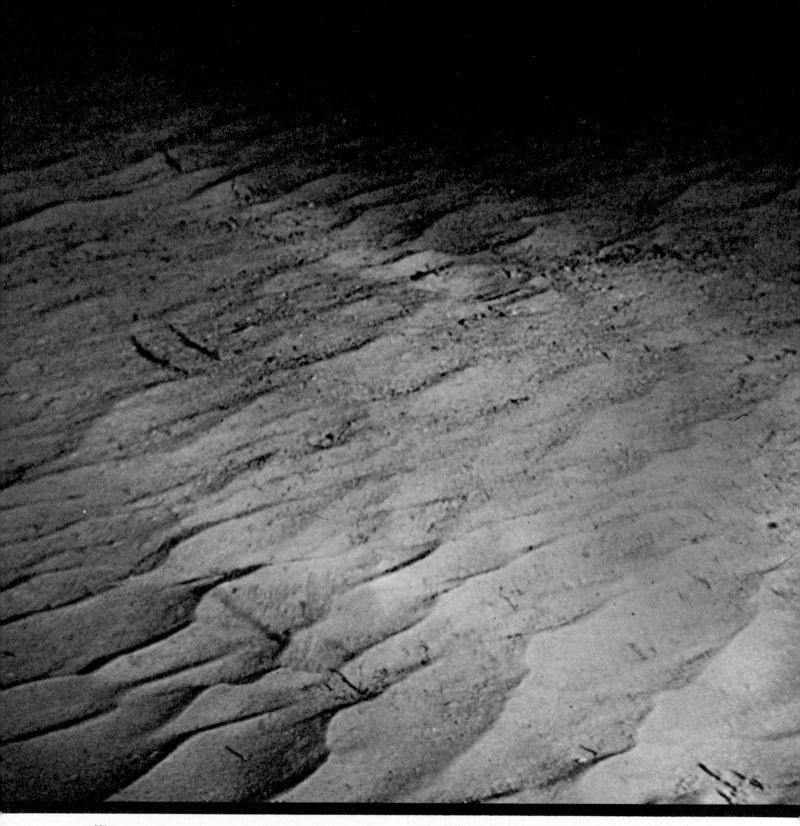

still warm. It gave support to those who believe in the theory of drifting continents. Molten lava flows from the scar of the Mid-Atlantic Ridge. The photograph above was also a surprise. It displays ripple marks at the base of a seamount in 7000 feet of depth. It proves that very deep currents can eventually be strong enough to carve such ripples—almost as if they were slow winds carving sand.

Camera Scooter

With his movie camera mounted on a motorized undersea scooter, this cameraman can slip and slide and race about chasing fishes, dolphins, and other divers. But if he isn't careful he could be wearing his face mask on his ear because the speed of the scooter in-

creases the force of the water against him. With his camera housed in a watertight plastic tube and held firmly on the scooter by a vibration-free mount, the cameraman has controls on the handles of the vehicle. He can move about freely with a minimum of effort. He also carries an underwater photographic light meter around his neck.

Problems of Filming

The aqualung freed us from the surface. But we still find ourselves tied to the surface for electrical power. Batteries work well for low power needs, but in filming with many powerful flood lights we require vast amounts of electricity. It was so in the past and it is still true. In these photographs it is plain what the problem is: bothersome cables streaming down from surface vessels or shore. In one effort to keep cables under control we tried attaching buoys to keep them floating off to the side of the work area. Eventually all facilities underwater must be free of surface encumbrances.

Filming a Dolphin

Underwater actors do a swim-by for the cameraman who is equipped with a 35-mm movie camera sealed in a metal cylindrical case. The underwater housing for the camera is only a refinement of much earlier cases made by undersea film pioneers. Although cameras in such housings are compact, addition of underwater lights to the unit makes it unwieldly. Lighting to best advantage and direction of subjects remain among the ma-

jor problems. Here, diver and dolphin pirouette past the cameraman. While most free dolphins don't readily submit to handling by people, this one apparently enjoys his work. Filming at close range as shown here, the cameraman is using a wide-angle lens.

"While most free dolphins don't readily submit to handling by people, this one apparently enjoys his work."

Stop-Action Photography

Stop-action photography was the technique used to capture this picture depicting a nudibranch moving through the water. To get the picture, the photographer planted his camera firmly and made a series of appropriate exposures at one-second intervals. The result shows the various stages of movement as the nudibranch, a shell-less snail, swings along.

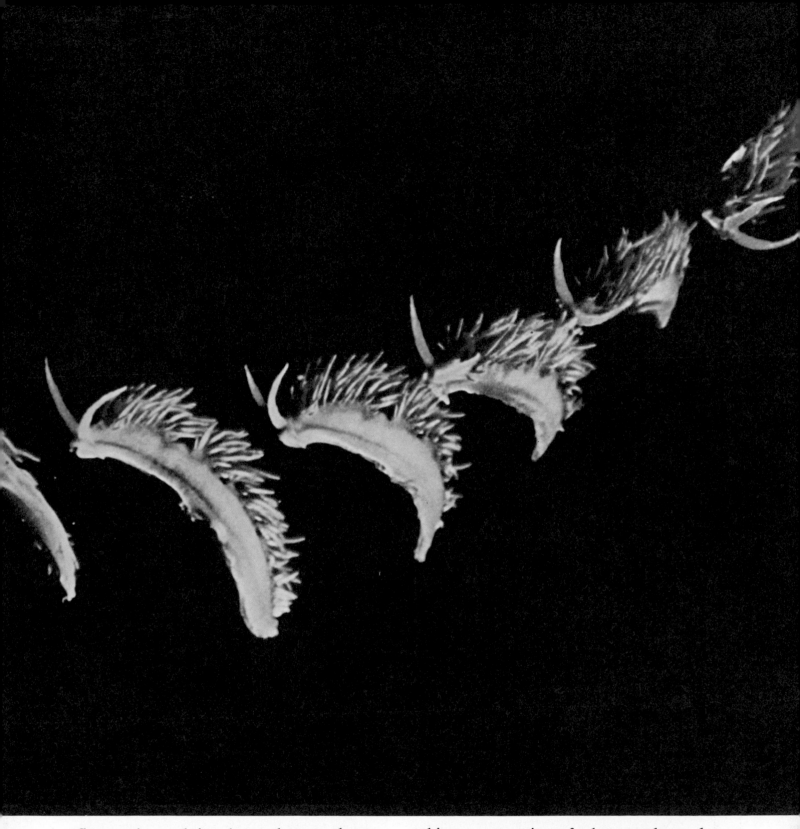

Stop-action and time-lapse photography are used to make such studies and to record the changes in a habitat over several hours', days', or weeks' time. They can be made either with a motor unit on a still camera taking a succession of photographs or by opening the shutter once and leaving it open while a strobe is triggered at the desired intervals. Through this technique we have learned much about the mysteries of motion.

Chapter IV. The Beam of Life

The source of all energy on earth is sunlight. It is the beam of life.

Without sunlight—even the meager amounts of sunlight that penetrate our atmosphere and our seas—green plants could not exist. Without green plants there would be no life as we know it.

How does the sun support the rich and abundant populations of living plants and animals? Through a complex and only partially understood process called photosynthesis, which involves the radiant energy of sunlight absorbed by the green pigment chlorophyll and converted into chemical energy. It is the chemical bonds which store this energy, and as these bonds are subsequently broken, energy is released, which can then be used for muscle contraction, growth and repair, and all other vital functions of the body. Initially solar energy is used to create more complex molecules from water and carbon dioxide. Water is split into hydrogen and oxygen; the hydrogen then reacts with carbon dioxide to form an organic molecule. In subsequent reactions more complex molecules are formed, such as sugars, lipids, and proteins of living tissue. Cells containing chlorophyll receive energy in visible light. Blue light waves contain more energy than other visible light. The amount of energy in light decreases with longer wave length so that X-ray waves carry the most energy, then blue light waves in the visible spectrum, and so on. Red carries the least energy in visible light.

Of the trillions of photons that reach earth, only a tiny proportion—perhaps 100 out of a million—are actually used in the process of photosynthesis. And of those 100, about half contribute to the growth of phytoplankton. Altogether, a not very efficient process. The oceans cover slightly more than 70 percent of the earth's surface —over 139 million square miles or 40 times the area of the United States. From that vast ocean area, enough oxygen is produced by the phytoplankton and other

> "Without sunlight—even the meager sunlight that penetrates our atmosphere and our seas— green plants could not exist. Without green plants there would be no life as we know it."

marine plants to support all the animals in the world.

If, however, the diatoms, flagellates, dinoflagellates, and coccolithophores—which make up the phytoplankton—and all the algae were no longer able to live in the sea, life would come to a grinding halt. The only way this might happen would be if the sun went out—a most unlikely occurrence that certainly would bring an end to life on earth.

If photosynthesis is so important to mankind, why don't we apply our vast technology to duplicate the process and assure super-supplies of oxygen? Actually, we have experimented with artificial photosynthesis—with only limited success. The process is not an easy one to duplicate. At best our attempts have been far less efficient than the natural process, which—as we have already seen—has its own limitations. Scientists have estimated that natural photosynthesis is about 37 percent efficient. Man's best efforts have been less than 10 percent efficient. Therefore, it is important that we not despoil the seas.

The Green Sea

Clear seawater appears blue. But add organic matter and a yellow pigment results. When the yellow of the organic matter mixes with the blue of the clear, nutrient-free water the result is green. Since much of the life that provides the yellow inhabits the shallow coastal waters of the tropics, there the sea appears green closer to shore, remaining blue farther offshore. In temperate waters, rich in nutrients, the green sea is a common sight.

Multicolored seashore. *At the shore off this island in the tropics the green tint of the water is evidence that the water is full of nutrients.*

Oceanographers, in making their various measurements, include the color of the sea as a factor. They use a comparator to determine the degrees of blue, yellow, and green of the waters they study. The richest fishing areas of the world, the Grand Banks off Newfoundland, Georges Bank off New England, and most waters covering the continental shelves, are green in color.

The Food Chain—Step One

These are the phytoplankton—the microscopic drifting plants in the sea that are the first link of the life chain. As a result of pho-

tosynthesis, tiny plants such as these produce trillions of tons of primary plant food
upon which all marine animals feed, either directly or indirectly. Requiring sunlight, these simple plants live in the top several hundred feet of the sea.

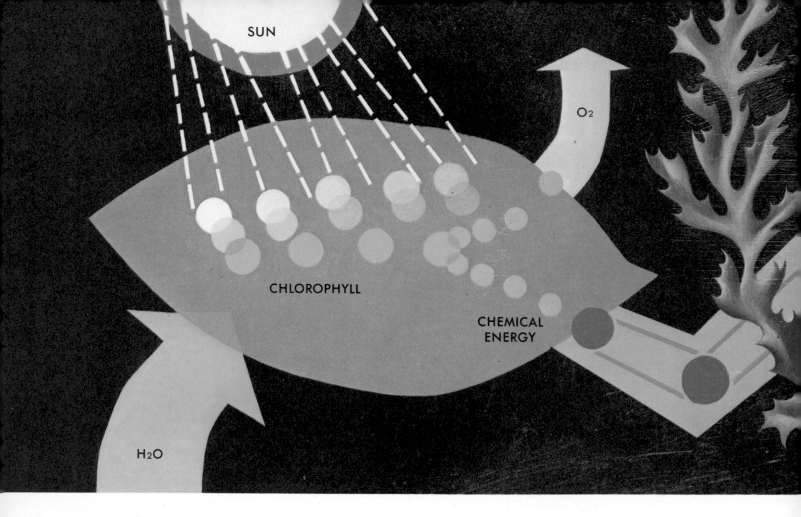

SUN

O₂

CHLOROPHYLL

CHEMICAL
ENERGY

H₂O

Photosynthesis

The sun radiates energy in several forms, including visible light. Those rays that come to earth must traverse more than 90 million miles of space, penetrate our atmosphere, and strike where they may on land and sea.

Sunlight that finds green plants—phytoplankton and seaweed, among others—initiates an amazing life-sustaining process, photosynthesis. As a result of this process, organic matter (the basis of food) is manufactured from inorganic matter, something man has been able to imitate only with great difficulty.

Photosynthesis begins when visible light is absorbed by the principal green plant pigment, chlorophyll, and other accessory pigments. This energy is utilized to split water (H_2O) into oxygen (O_2) and active hydro-

gen (H). The oxygen is liberated into the atmosphere where it may be subsequently respired by other organisms. The hydrogen is combined with carbon dioxide (CO_2) to form organic molecules, figuratively expressed as (CH_2O), and more water. The organic molecules may be later rearranged, combined, modified, and coupled with other compounds to produce the broad variety of substances found in living organisms, that is, lipides, carbohydrates, and proteins. These fuel the growth of seaweeds, diatoms, and other plants, which in turn feed either directly or indirectly the ocean's animals. Ultimately these organic molecules are reoxidized through respiration in animals and plants to water and carbon dioxide with the release of energy. Then they are recycled in the biosphere. In this way, the process of life is ultimately driven by the sun's energy.

Thus, the plants provide not only a balance

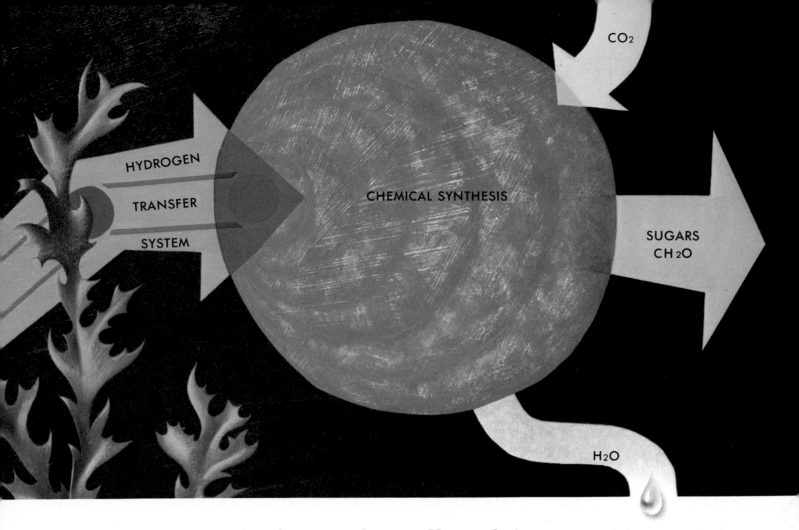

HYDROGEN
TRANSFER
SYSTEM

CHEMICAL SYNTHESIS

CO$_2$

SUGARS
CH$_2$O

H$_2$O

against our consumption of oxygen and our production of carbon dioxide, but provide us with nutrients and other materials we need to utilize our various plant and animal foods —sugars, starches, and enzymes to name a few. In the long run the animal matter we consume depends on plant life too.

Because plant life in the sea needs the sun's radiant energy to start this complex chain of

> "The sun's rays that come to earth must traverse more than 90 million miles of space, penetrate our atmosphere, and fall where they may on land or sea."

events, and because the sun's rays penetrate in any significant amount only 200 or 300 feet, it is in those surface waters that the greatest amount of activity takes place. And

Photosynthesis. *The result of this process is the biosynthesis of sugars which are needed for the growth of seaweed and other plants. This is accomplished by splitting of water into oxygen and hydrogen. The hydrogen reacts with carbon dioxide to form organic molecules.*

since so many animals feed directly or indirectly on those algae, that is where most of the life in the sea is found. There are exceptions. For some algae, unexplainedly, are found in greater depths. Did they drift down? Were they carried there? Or do they occur naturally in those greater depths? No one knows for sure. But it seems evident that little or no photosynthesis can take place in those algae that live so deep the sun's rays do not touch them with its energy. There are animals living in those greater depths also. But these are not algae-eating fishes and invertebrates. Rather they are animals that feed on organisms that feed on the algae of shallower waters.

Natural Photosynthesis

Proof that algae manufacture oxygen in the process of photosynthesis is seen in this photograph of tiny oxygen bubbles on the green algae covering a submerged rock. The bubbles form if the gas is produced faster than it can dissolve in the seawater. When the bubbles get large enough to break away from the tips of the plants, they float to the surface. Oxygen not absorbed by the water is given off into the atmosphere. Algae in the ocean may manufacture 50 to 70 percent of the atmospheric oxygen we breathe.

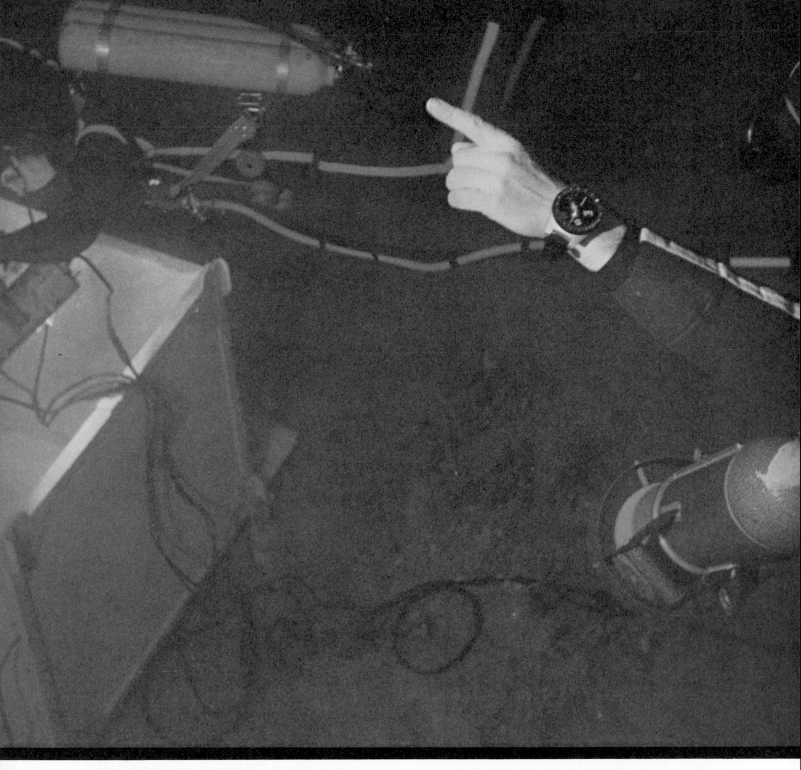

Simulated Photosynthesis

More than 300 feet beneath the surface, in virtually lightless depths of the Mediterranean off Cap Ferrat, France, oceanauts of Conshelf III experiment with simulating photosynthesis. At that depth, because of the lack of sunlight, natural photosynthesis does not occur in any significant amount. But by using artificial lights our oceanauts appear to have met with success. Increasing the productivity of the oceans by introducing the energy of light in the deeps could boost the food productivity of the sea.

▲ A

Various Algae

Sunlight, water, carbon dioxide, dissolved nitrates and phosphates, and minute amounts of elements are the ingredients algae and marine grasses use in the process of photosynthesis. Along a rocky California shore, brown algae float in the shallows.

A / Brown algae. In the photograph above, a type of brown algae is seen, treelike, with the sun shining on it.

B / Udotea. This is actually one of the green algae, although it appears to be a different, luminescent shade here.

C / California shoreline. Both green sea grasses and brown algae thrive along the reefs of Carpinteria, California.

Chapter V. Pulse of the Ocean

A heartbeat of majestic proportions is recorded by echo sounders on ships dispersed over the seas—a slow, daily universal vertical migration: the rise of the deep scattering layer to the surface as sun's light fails, its return to the depths before dawn.

The DSL is only the echo sounder's revelation of nomadic populations, one of the ocean's mobile "provinces." In lighted surface waters we encounter huge clouds of marine life separated by great expanses of

> "Near the surface, in the upper 300 feet, the sea may be turbid with life. Here, triggered by sunlight, photosynthesis occurs. Planktonic plants and animals abound. Beneath this relatively thin layer of living matter, in ... diminishing sunlight, is a vast expanse of water sparsely populated with life."

apparent wasteland. The deeper one explores, the rarer life becomes. But near the surface, in the upper 300 feet, the sea may be turbid with life. Here, triggered by sunlight, photosynthesis occurs; planktonic plants and animals abound. Beneath this relatively thin layer of living matter, in a twilight of diminishing sunlight, is a vast expanse of water sparsely populated with life. This desert stretches to a depth of about 1500 feet and contains clear water of great transparency, although natural vision is limited by the absence of light. Beneath the barren layer is another tier of living sea in which William Beebe described animals "as thick as I have ever seen them."

Many attempts have been made to photograph or observe the creatures of the DSL.

They have been largely unsuccessful. As cameras are lowered toward them the creatures flee before they can be photographed, perhaps warned by lights accompanying the cameras, the whirring of motors, or pressure waves. Beams of the bright lights on our deep submersibles cause them to take flight. When a camera is lowered, it can be observed on an echogram approaching the DSL. As it closes in on the animals, the layer scatters. And as the camera continues downward, the DSL re-forms above it. Some of the inhabitants can be recorded on film.

The layers of the ocean are not fixed to a single level, but the creatures comprising each cloud of life tend to remain in distinct groups. Those of the DSL rise to the surface and descend to the depths, responding, probably indirectly, to the light of sun and moon. As the sun drops below the horizon and the waters darken with the coming of night, deep dwellers respond to the decreasing sunlight and begin their long vertical migrations toward the surface. This is not to say that the entire DSL suddenly rises en masse, but that each species, in its own time, ascends. The animals at the upper side of the DSL may rise only a few hundred feet to reach the surface, while squids, spending their daylight hours at the lower limit, may climb thousands of feet.

In the darkness of the deep waters they inhabit during the day, the residents of the DSL are relatively safe from predators. At night they can rise, still relatively safe from visual detection, and feed in greater safety than if the waters were lighted. Also, in the process of photosynthesis phytoplankton may release toxic substances. By remaining below the photic zone at a depth where photosynthesis can't occur, DSL inhabitants avoid contact with these substances.

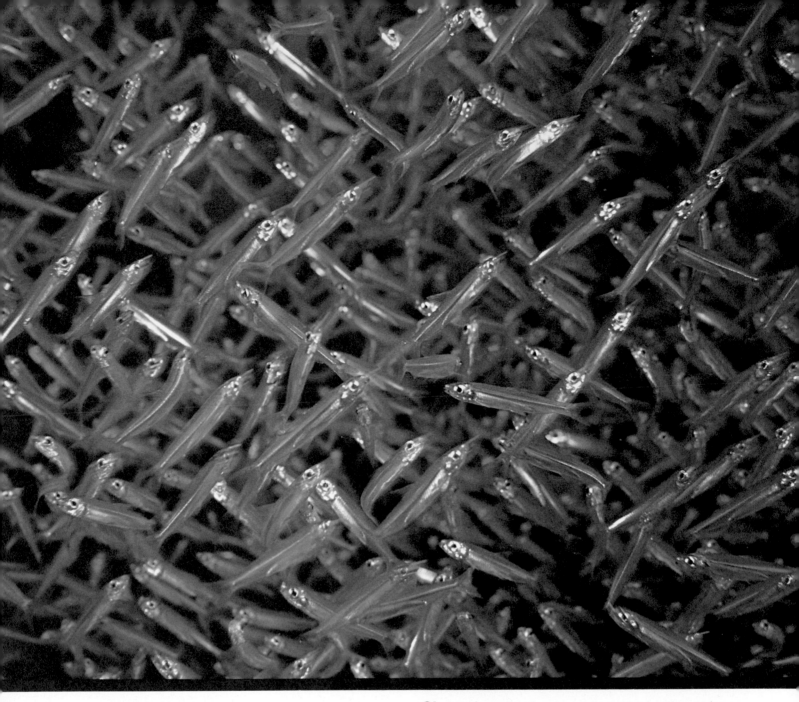

Jacklighting

Jacklighting is using a powerful beam of light to freeze animals in their tracks in order to hunt them at night. Since antiquity fishermen have used light to capture fish. The method is very common around Africa and the Mediterranean Basin. Still today many fishermen use a devastating form of jacklighting. While a light hung over the side of a boat frightens some species away, it attracts others, large and small. Hovering just

Glass minnows. *In the picture above, glass minnows are mesmerized by a beam of light. This process is called "jacklighting" and is a method used by fishermen.*

beneath the surface they bask in the glow like insects around a light bulb on a warm summer's night. Predators gather around the area. A diver emerging from one of these light-crazed crowds reported, "It was incredible. A hundred of them would be hitting your body at the same time. You felt like a piece of paper in a typewriter."

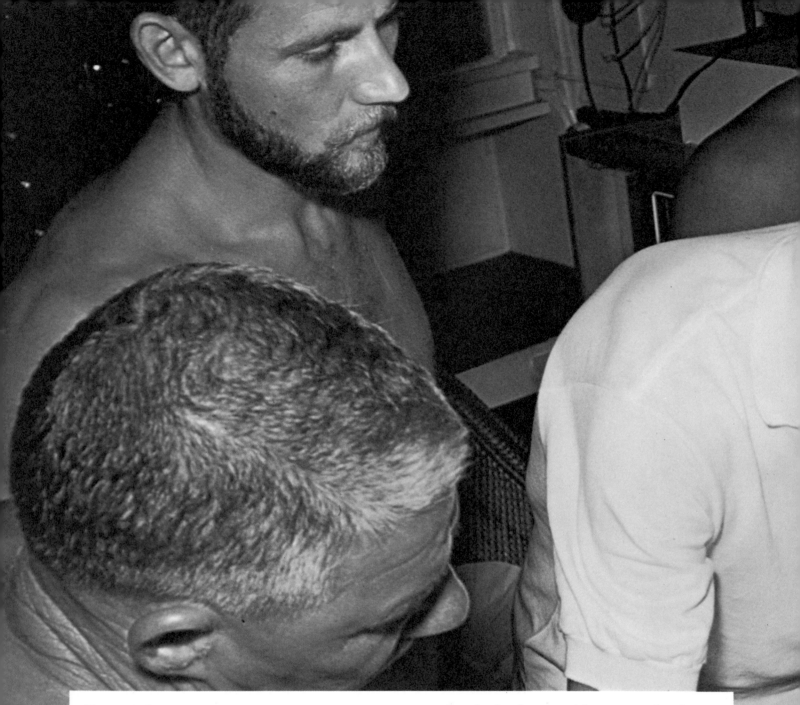

Deep Scattering Layer

As the sun sets and we plunge into the darkness of night, a changing of the guard takes place among the animals of the sea. The deep scattering layer (DSL) is stimulated to rise toward the surface from their daytime homes in the depths. Our first efforts to learn more about this phenomenon began in 1953 when we used a camera system designed by Harold Edgerton. The apparatus consisted of a frame with two steel tubes—one containing a camera, the other flash equipment. The tubes were angled toward one another so that the flash would reveal anything six feet from the camera. From *Calypso*'s deck we could follow the explosions of the flash to a depth of 500 feet.

Echogram. *Here Captain Cousteau and members of his crew study an echogram on which deep scattering layers can be spotted above the bottom profile.*

Appetite Satisfiers

In winter and spring Alaskan fur seals live in cold northern Pacific waters, only occasionally in groups or pairs. Since they have been protected against overzealous hunters by international agreement, these animals have made a remarkable population recovery. It is estimated that as many as 4 million of them come ashore to breed each year on the Pribilof and other islands. Fur seals are hearty eaters. They need a lot of calories to

maintain their body heat in icy waters. If they eat as much in the wild as in captivity, it is likely that a million of them would eat a billion pounds of food annually. Where does all this food come from? Now some specialists believe their diets are comprised

Weddell seal. *The male of this species can weigh close to 700 pounds.*

substantially of lanternfish, those silvery little creatures that rise each night as part of the deep scattering layer.

Squids and Their Pursuers

Photographs of the deep scattering layer have revealed squids fleeing from the probing eyes of our cameras and bright lights. On 1000-foot dives in the diving saucer I have observed young squids, only several inches in length, in flight ahead of us, probably blinded by the lights, darting to and fro, rushing away above us and then returning.

Squids are normally shy creatures, retiring to the darkness of the depths by day, protected there from the rays of the sun, and rising at night to the surface. Some squids are attracted to the lights fishermen hang over the sides of their boats on dark nights, or perhaps are attracted to the other tiny creatures that gather around the lights. When a light is lowered into the water, or suspended above it, fish and crustaceans

swarm beneath. Squids take station at the perimeter of the illuminated area and dart across it, snatching up delicacies during their brief passage. At the same time squids are prey for other animals. On one occasion *Calypso* happened to be in an area where the mating ritual of small opalescent squids was taking place. Here the squids rose in great numbers to the surface during the night, fertilizing and laying eggs. Our divers observed

Attracted by light. *Either attracted by light from the boat, or by the other creatures attracted to that light, squids (left) swarm around our boat at night.*

Attracted by squid. *As usual, the squids were in evidence around the boat at night. The photograph (above) was taken when hungry sharks noticed them.*

the activity carefully, but sharks in the area were not as interested in watching as they were in eating.

Nocturnal and Diurnal Color

The same Caribbean snapper looks different in the daytime than at night. Color and pattern changes take place as darkness approaches. The fish swims actively during daytime and sleeps on the bottom at night. The changes presumably are a defensive adaptation. During the day, when swimming in midwater, it is colored a fairly even red, perhaps a little darker above than below. The red appears only as a dark color, blending the fish into the background of the 100-foot depths where it swims, because red light has been eliminated by absorption in the uppermost 30 feet. At dark the fish descends to the bottom where it lies at rest for the night. There it adopts a blotchy red coloration with light patches, blending with the uneven colors of the sea floor.

Mesmerized Fish

Concentrated beams of light hold many species of fish mesmerized. Here a triggerfish stares transfixed into an underwater lantern. Fish virtually hypnotized by lights will frequently allow themselves to be hand-picked by divers. Others come out of their trances when touched and dart away into the safety of the darkness. One night at Port Calypso, our underwater archaeological digs at Grand Congloué Island in the Mediterranean, our divers came across an enormous school of horse mackerel. The fish became entranced by our lights. Hundreds upon hundreds of these silvery fish swam crazily into and around the divers. Later, when the men climbed aboard the *Calypso*, one diver danced wildly about until we pulled a mesmerized fish from inside his diving suit.

Chapter VI. Light Alive

Before he mastered the art of sailing, man rowed his vessels to sea. As his boat glided through the sullen, black nighttime waters, the early seaman saw strange and mysterious stirrings in his wake. Wherever the water was disturbed by the passing of the boat or the stroke of an oar flashes of light occurred, flaring brightly for a brief moment.

Men wondered for ages about these "underwater fireflies." But a number of other marine animals also bring light to the dark waters of the seas. Luminescent animals are found in nearly all ocean areas, from shallow coastal seas to the deep waters of the abyss. Little sunlight penetrates the waters beyond 600 feet, and below 2000 feet almost none can be detected. The luminescent creatures that live there provide the only visible light. About three-fourths of all deep-ocean species, including 80 to 90 percent of the total population, are able to produce light, directly or indirectly.

Many animals have unique cells called photophores containing specialized light-gener-

> "About three-fourths of all deep-ocean species, including 80 to 90 percent of the total population, are able to produce light, directly or indirectly."

ating tissues. They vary in size, shape, color, and location on the body. One abyssal fish (*Malacosteus*) proudly displays reddish lights on portside and greenish lights on starboard. The light-producing cell may be isolated from other parts of the body by a cloak of black pigment; reflective tissue may line the photophore. In other cases living light is produced not by the animal that exhibits it but by bacteria in pockets on the surface of the host's body. These bacteria glow continuously. Some hosts have developed mechanical means to darken themselves—flaps of skin that can be pulled over the pocket of light like window shades. In some other fish the scale covering the pocket is thickened at the center and forms a lens that focuses the light and increases its effect. The position of the pockets the bacteria occupy varies, and males and females of a single species may even have different lighting patterns.

In a flash photograph tiny marine organisms seem to be a lifeless mass, but in the black of night they give off an eerie blue light. Luminescent creatures may be large or small, from shallow water or deep, but they have one thing in common: their light is nearly 100 percent efficient. Most of the energy used to produce it is converted to light; in contrast to our electric bulbs, little heat is given off. The light is usually blue, and is about the same wavelength as the blue in sunlight, penetrating farthest into the sea.

Animals dwelling in the eternal night several thousand feet down have large, sensitive eyes capable of detecting light much fainter than we can see, and those with lights take advantage of this sensitive vision. Since the lighting display varies between species the patterns may assist in recognizing members of the same species, or members of the opposite sex. The pattern of the signal lights may spell mutual attraction for mates or mutual repulsion for members of the same sex.

Luminescence. The top photograph was taken using only the light of the bioluminescent plankton on the man's hand. The bottom photograph shows the same plankton under artificial light.

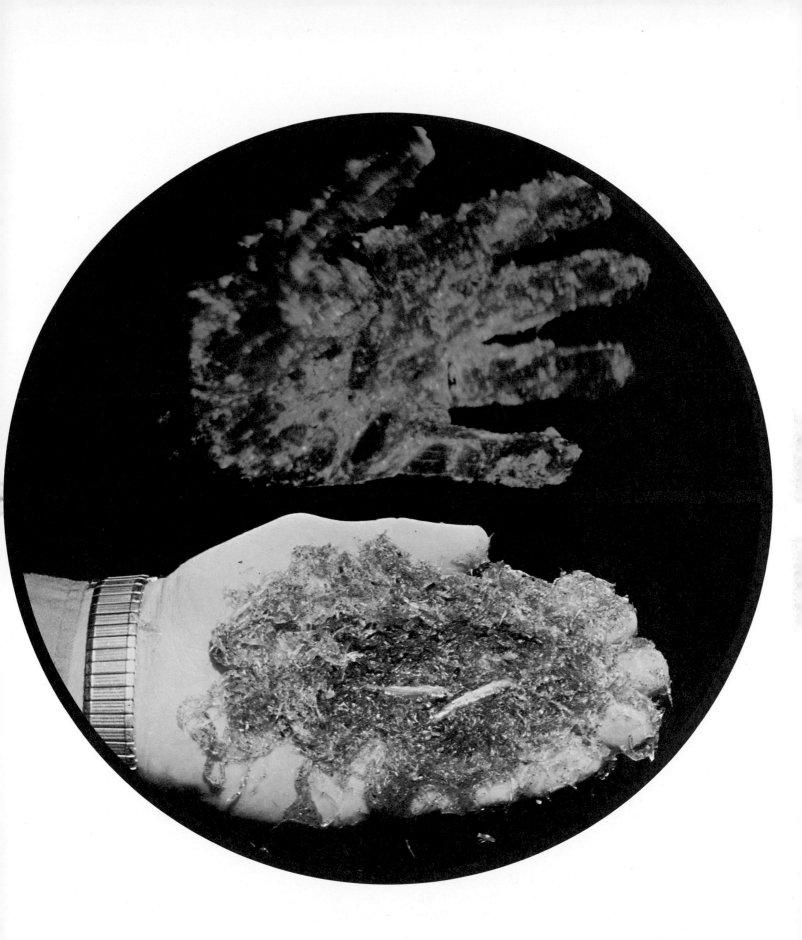

▲ A

Bioluminescent Deep-Sea Fish

Bioluminescence is a characteristic common to many creatures of the deep sea.

It has been estimated that as many as 75 percent of deep-dwelling species, and perhaps 80 percent of the total deep-sea population, are luminescent, including some worms and sea fans as well as fish. Luminescence differs from species to species, each having its own distinctive patterns and colors. The luminous organs of fish, whether photophores or envelopes with luminous bacteria, are located in different areas of their owners' bodies. Within a species the location of these organs is uniform, with all members of the same sex resembling their relatives. Some

have lights on their heads, others near their tails. Some have luminous fins, while in others the light organs line the sides of their bodies. In some species the luminous organs are located inside the mouth!

There are over 200 species of lanternfish, each a miniature constellation. The lights are generally round, looking like pearl buttons, probably used primarily as signaling devices. Located along the sides of the fish, they are probably of little use in lighting up the water sufficiently to allow the fish to see. Some lanternfish have luminous organs on their tongues.

The first ray of the spiny dorsal fin of female deep-sea anglerfish, living in perpetual darkness at depths of 6000 feet, has modified

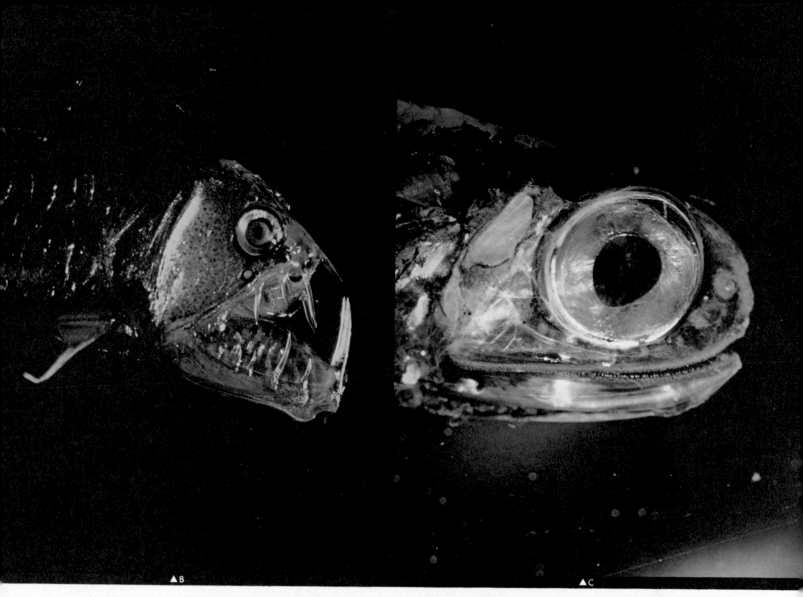

and migrated to the front of the fish's head; from it hangs a fleshy bulb used as bait. The device is complex, usually with multiple lighted filaments to entice prey closer. There is a great amount of variation in the characteristics of the luminous lure. Immature males, with tubular eyes and good vision, probably recognize the variations in the lights and identify mates by them. Since some males parasitize their mates, living off the female's body fluids, they have no need for these fishing lures and do not have the apparatus.

The viperfish has a fishing lure, too, but his extends from the second ray of the dorsal fin. Equipped with a luminous tip, the elongated ray dangles in front of the mouth of a hunting viperfish. The viperfish has about

A / Anglerfish. This fish, as we can see in the photograph at left, has a luminous "bait" which dangles in front of its mouth. There are several species of anglerfish, ranging from six to 40 inches long.

B / Viperfish. This nasty-looking fish has 350 photophores on the roof of its mouth with which to attract prey.

C / Lanternfish. These fish have many photophores lining the sides of their bodies. Their brightness and light patterns vary from species to species.

350 photophores on the roof of its mouth, and another battery lining the lower part of its body. These may attract the small fish and shrimp upon which the viperfish feeds. It can be said the viperfish feeds as it breathes, since by opening its mouth to admit water for respiration it reveals the photophores inside to the curious ones.

The Squid's White Ink

Once, at 3300 feet below the surface, I peered out of the bathyscaphe *FNRS-3* and saw something like this deep-sea squid. As I watched it shooting past I saw it squirt a cloud of phosphorescent ink. At first, in the gloom of the ocean depths, I thought it was white ink. And I said to Commander Houot, my companion in the bathyscaphe, "A squid making white ink!" "You have strained your eyes," Houot replied. (We both knew that squids have brown ink). Another squid flashed past the window leaving a white cloud and I realized it glowed with phosphorescence. "Swish. Another luminous puff," I cried out again. Houot said, "Let me have a look." And soon he too saw the puffs and clouds of ink that glowed in the otherwise all-pervading darkness. Squids have evolved this ability to eject luminous ink in order to distract predators, who will watch the puff while the animal jets away—fooling his enemies.

Bioluminescence in the Shallows

In dark, murky waters, a disturbed sea pansy glows with a faint bluish light. There is no known functional significance for their bioluminescence. Sea pansies are relatives of

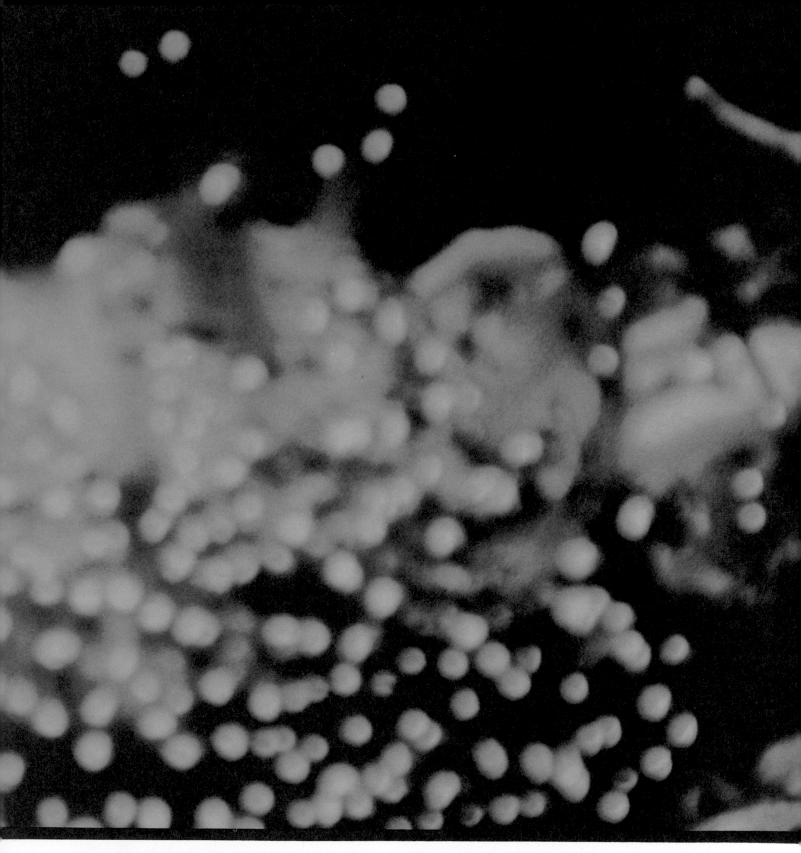

sea pens and corals, sea anemones, and jelly-fish. They fasten themselves to the bottom by stalks and patiently await the arrival of food drifting by. The food, zooplankton, gets stuck on the polyps which cover the top surface of the sea pansy itself. When a diver touches a sea pansy, a series of beautiful radiating circles of light results.

Flash of Mirrors

Fish that have silvery or gray sides are to be found practically at all depths, from the sunlit surface areas to the darkest abyssal provinces. Though conspicuous under certain lighting conditions, the silver is a form of camouflage, making it difficult for predators to see these fish. Sunlight penetrating the water and striking the silver sides of these animals is reflected downward, as though bounced off a mirror. An observer below silvery fish sees this reflected light as if it were light shining through the background surface waters above the fish. When fish so colored are below the observer, the light is

reflected into the depths, and goes unseen from above. Since the backs of the creatures are dark, they blend into the background waters.

Schools of shiny fish present a difficult situation for predatory species. When attacked the school disperses instantly—small bodies scattering in every direction. Unless a preda-

Menhaden. *Above, a school of immature menhaden—in hundreds or even thousands—is seen with light reflecting mirrorlike off the bodies. These fish are not bioluminescent, but natural light acts on them in such a way as to confuse predators.*

tor has singled out one individual, it is confused by the flashing bodies. While these fish do not produce their own light, their ability to reflect light is essential to their survival.

95

Chapter VII. Prometheus

According to Greek myth a Titan named Prometheus stole fire from the heavens and brought it to earth. Eventually Zeus caught and punished him. But Prometheus had given mankind fire and thus artificial light.

The French marine biologist Louis Boutan was a sort of Prometheus of underwater photography. After he made the first successful underwater photographs, he devised the first working underwater flash. In his own words he described the device thus: "The first (photographic) lamp consisted of a spiral wire of magnesium in a glass balloon containing oxygen and a fine platinum wire connected to the two poles of a battery. When the current was turned on, the wire reddened and ignited the magnesium which oxidized with a brilliant light."

The pioneer of underwater color photography, the marine biologist W. H. Longley, may have been less imaginative but he was more spectacular. He used a pound of magnesium powder on an open-bottom raft floating above his intended subject. A reflector over the raft diverted light downward when it was ignited by a topside assistant. The resulting explosion gave off an equivalent of 2400 flashbulbs' worth of light.

The next step was the development of the standard flashgun and bulbs for underwater use. Some photographers still use this lighting system to good effect. The disadvantages are the buoyancy of the bulbs and the danger of implosion of the bulbs under pressure. Luis Marden, now chief of *National Geographic Magazine*'s foreign editorial staff, adopted a butcher's chain-mail glove after being gashed by an imploding bulb.

Harold Edgerton's strobe was used for undersea work. Encased in a pressureproof and waterproof case, the electronic flash was adapted by bringing the color temperature down close to that of sunlight. Its use is not

> **"W. H. Longley used a pound of magnesium powder on an open-bottom raft floating above his intended subject. A reflector over the raft diverted light downward when it was ignited by a topside assistant. The resulting explosion gave off an equivalent of 2400 flashbulbs' worth of light."**

restricted to still photography, but it is used for stop-action and time-lapse motion pictures to study, for example, the movement of sea urchins and sand dollars.

Floodlights, of course, have been waterproofed and taken below too. In a particularly dramatic application, Albert Falco, chief diver aboard *Calypso,* once beamed a hard, sharp sliver of light at a silver bream while diving near Gibraltar. The light held the fish transfixed.

Some marine animals seemingly create their own light. A long slender sponge appears yellow at 110 feet. Several species of coral display red at the same depth, long after all red and yellow light from the sun has been filtered out by the water. Brought to the surface these organisms are a dull, mud brown. The answer to this perplexing sight: fluorescence.

Diver and flare. The flare with its thick stream of bubbles is beautiful in itself, but its use as a light source is somewhat limited. The brilliantly colored grouper in the foreground would have shown up only as a dark silhouette had the photographer not used other lighting on it.

Posing. *In the drawing above, Boutan, in diving gear, readies his bulky flash equipment for an underwater photograph.*

Self-portrait. *Boutan took the first underwater photograph in 1893. The self-portrait at right was taken a few years later.*

Boutan Lights

Clad in fashionable ankle-length striped trunks, Louis Boutan, the father of underwater photography, poses heroically in the self-portrait at right. He took this picture of himself 12 feet below the surface.

Besides establishing undersea photography as an invaluable tool for the marine scientist and others, Boutan brought light to the submarine world. In the illustration above Boutan, garbed in a diver's dress, prepares to set off a flash to take a picture of fish against a white placard he placed there as contrasting background for his subjects. In this system, an alcohol lamp in a bell jar was fixed firmly to the top of a barrel and connected to it by

an airway. On one side of the barrel a waterproof case of highly inflammable magnesium powder was connected to the inside of the bell jar and aimed at the flame of the lamp. With a bulb to squeeze the magnesium powder out onto the flame, the diver-photographer could make a larger or smaller flash depending on how he squeezed the bulb. Air in the barrel enabled the lamp to burn for many minutes. As long as there were air and magnesium, he could take flash pictures, for he had already devised a means of changing photographic plates underwater.

Another means of underwater lighting that Boutan devised was to enclose a pair of arc lights in heavy metal and glass spheres which were fixed to either side of a camera.

Underwater Color Photos

William H. Longley pioneered underwater color photography with a bang. He used a full pound of magnesium powder which went off with an enormous explosion to illuminate the first color pictures in July 1926. They were published in the January 1927 *National Geographic Magazine*. Longley took the pictures in the waters off the Dry Tortugas beyond the Florida Keys with Charles Martin, at the time the head of *Geographic's* photo lab. Martin had to invent all sorts of devices for this pioneering effort. And they worked. Longley carried the eight pictures to *Geographic's* Washington offices and announced, "Herewith are eight autochromes of genuine submarine life; the first ever

taken." But no more underwater color photographs were published for 20 years—not until, with the aid of the aqualung and flash equipment, Jacques Ertaud and I began shooting pictures in the Mediterranean. Since then millions of color pictures have been taken underwater, many with the aid of a flash which enables photographers to capture the colors of life in the sea in more

Everyday scene. Taking underwater color photographs is no longer a novelty. The diver/photographer seen here is taking a picture of a Nassau grouper with today's underwater equipment.

vivid detail than can be seen with the naked eye. The diver shown here is using an ordinary underwater flashgun with his camera in a watertight housing. What was a rarity 15 years ago is today a commonplace.

Stroboscopic flash. Well-equipped photographer-diver taking close-up of sponges lighted by a stroboscopic flash.

Stroboscope

When Harold Edgerton invented the stroboscope in 1933, it opened the way to ultra-high-speed photography. The strobe was designed as an instrument for studying motion with extremely rapid, brilliant, short-duration flashes of light—so short-duration that Edgerton could make stop-action photographs of bullets as they entered apples, hammers smashing light bulbs, the movements of an eye as it scanned a page. It's doubtful the inventor had in mind any underwater application for this light measured

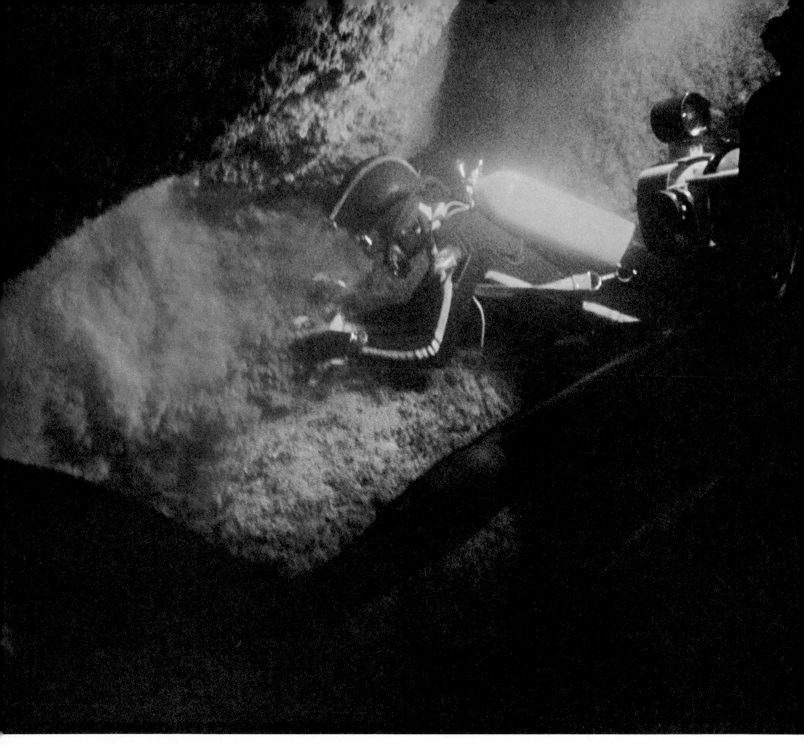

Aiding sunlight. *A diver (center) is filmed by the cine camera (at right) while the cave is lit by flood-lights.*

in ten-thousandths of a second. But when we met on board *Calypso,* and the problem of underwater lighting was posed, Papa Flash, as we call him, had answers. The strobe eliminated the need to carry flashbulbs.

And it provided an extremely strong light source. But it required a waterproof housing.

The problem was to seal the circuits so that the photographer wouldn't be electrocuted as he triggered the strobe or electronic flash. Seawater is an excellent conductor of electricity and even a slight leak of power from the strobe's batteries could have serious consequences.

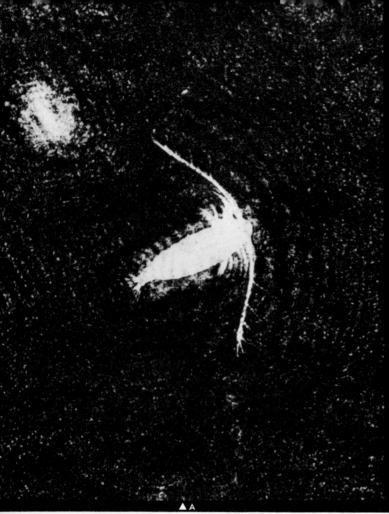

▲ A

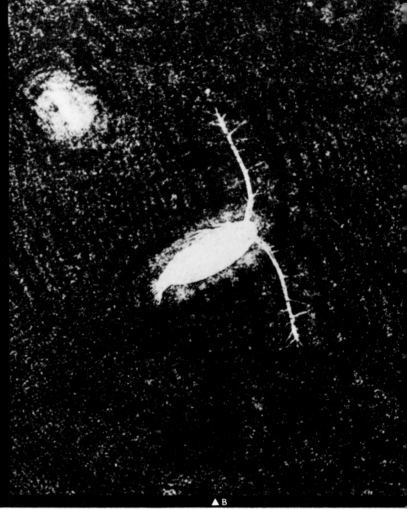

▲ B

A, B, C, D | Hologram. These are photographs of a tiny copepod. Were we able to reproduce the hologram itself here, the copepod would be seen as three-dimensional.

E | Photo multiplier. This device allows the viewer to see in almost total darkness. It can be attached to a camera so that a photograph can be taken in an unlighted scene.

Sophisticated Photography

The laser beam makes it possible to take three-dimensional photographs and motion pictures. The art is called "holography," a word that is appropriated from the Greek meaning "to write all." The photographs on these pages were made from holograms taken of a tiny copepod. If you were to see the hologram from which they were made, you would be able to see the little animal in the round, as if you were looking at him in a tank of clear water.

Holograms are made by directing parallel beams of pure laser light at a three-dimensional object and at a mirror. The rays of laser light reflected from the mirror and

those bouncing off the object interfere with each other and set up small patterns which can be recorded directly on film. Processed, the film will look blank until it is placed in laser light where the original wave patterns will be reconstructed.

Another remarkable device is the "owl eye," basically a photomultiplier, a mechanism so sensitive it is able to give a picture with only the faintest amount of ambient light. If the owl eye receives one photon of light, it can multiply it to 10,000 or even 80,000 photons and enable us to see in the dark as if it were daylight. But it must receive one photon to be activated. To supply this, we plan to equip the diving saucer with a 1-watt bulb placed at the rear of the submarine.

▲C ▲D ▼E

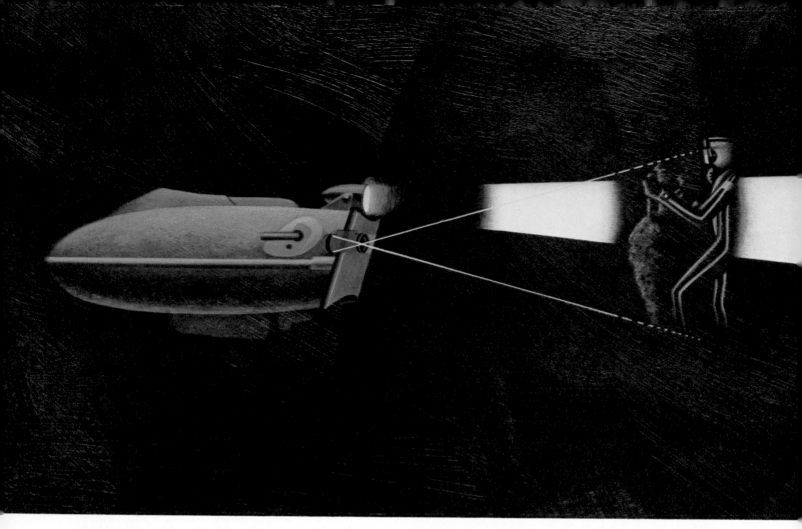

Mirage Photo System

Light reflected and scattered by particles in seawater creates a milky screen between

> "A sophisticated system has been devised to reduce the amount of light scattered and to permit illumination of an object as far as 40 meters away, an unheard-of feat underwater. The system works by sending out an ultrashort pulse (say, 1/100th of 1 millionth of a second) of bright light that travels to the subject, illuminates it, and reflects back to the camera."

the light source and the object to be viewed or photographed. It degrades the quality of

the image perceived and drastically shrinks the range of visibility. A sophisticated system has been devised to reduce this backscatter and to permit illumination of an object as far as 40 meters away, an unheard-of feat underwater. The system works by sending out an ultrashort pulse (say 1/100th of 1 millionth of a second) of bright light that travels to the subject, illuminates it, and reflects back to the camera. In traversing the distance twice (out and back) much of the light is diffused and builds a luminous wall between the eyes and the object.

To eliminate most of these effects, the viewing or photographic porthole is fitted with a "shutter" made of a special crystal containing metal particles. When it is subjected to a magnetic field, the opaque particles within the crystal are oriented in such a way as to allow light to pass through. The alternation

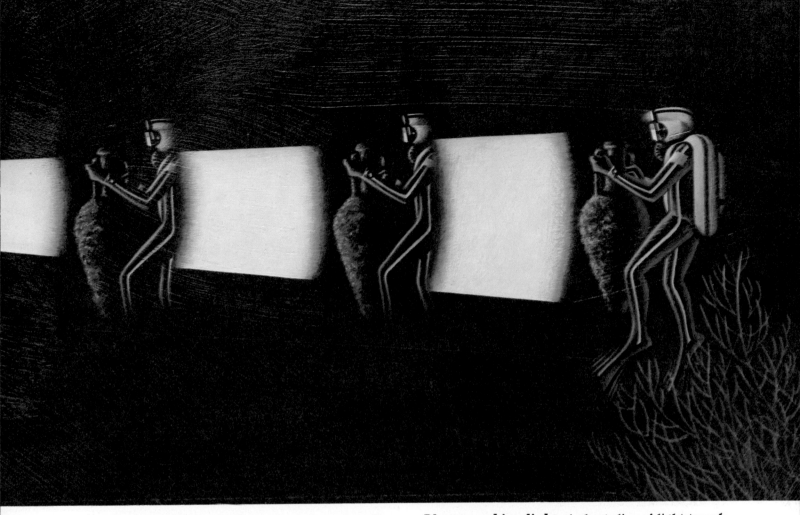

Photographing light. A short slice of light travels to the diver and is reflected from him. It travels back to the camera and activates the shutter which only opens at the exact moment pure reflected light reaches it.

between opaque and nonopaque in the transmitting properties of the crystal can take place extremely rapidly, making the crystal an ultrarapid magnetic shutter. As a result, only a small 10-foot-thick slice of the original beam can be received which will contain all the information about the area to be observed, but a minimum of scattered light. This remarkable Mirage system will allow scientists to study the depths with longer and clearer vision.

Such a system installed on the diving saucer would give the pilot and observer vision at a much greater distance than they have at the present time. Coming upon an object of interest, they could position their vehicle and photograph the slice of light reflecting from the object almost clean of scattered light.

The Mirage system will require a complex, computerized range finder to work in con-junction with an extremely sensitive timing device triggered as the light pulse is activated. One control would choose the distance at which the pilot wishes to observe or photograph, another control would determine the thickness of the slice of water to be illuminated. As many as 1 million pulses of light per second could be generated without interference. But 100 pulses per second would be more than ample to take advantage of the persistence of images in the human eye's retina.

Because this system uses techniques that are still at the very edge of modern possibilities, the cost of perfecting it may delay or prohibit its practical development.

107

Fluorescing algae. *Above, a type of algae is seen fluorescing. Some marine organisms fluoresce when placed under an ultraviolet light, others have the ability to do so in the presence of natural light.*

Fluorescence

Under certain conditions otherwise drab marine organisms glow with splendor. An ordinary algae or piece of gray coral placed under an ultraviolet light becomes a shimmering gem. What magic is this?

Coral has the ability to fluoresce, to emit light of a color different than that of the external source; it is sensitive to a particular part of the spectrum of light. Other fluors (substances having the ability to fluoresce) are stimulated by pressure, heat, or X rays. Fluorescence occurs when energy is absorbed by an atom or molecule, forcing electrons spinning around the nucleus or nuclei into an orbit of higher energy; when these electrons return to their original orbit, ener-

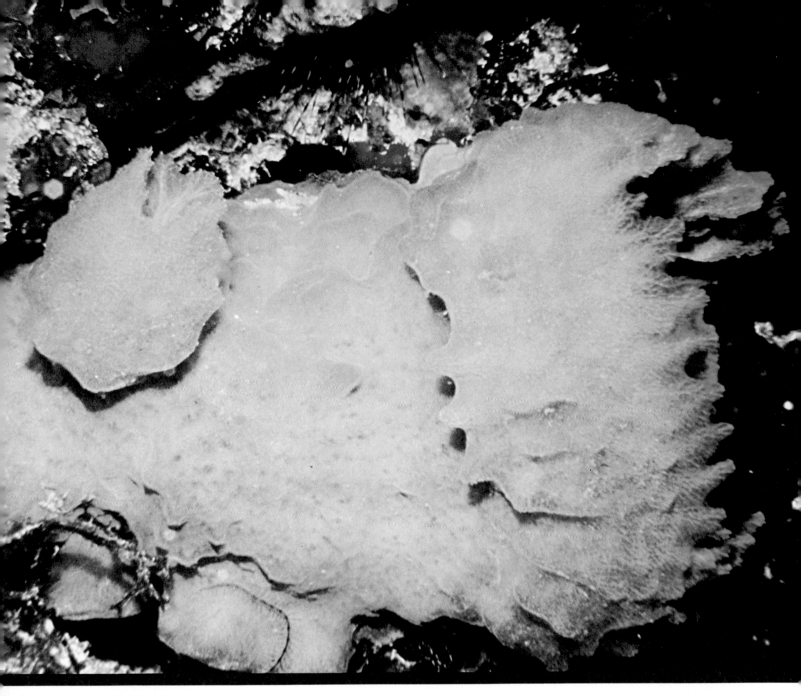

Glowing sponge. The beautiful sponge, above, is the color of a fluorescent light bulb with yellow overtones. In its nonfluorescent state, however, it could be quite drab and dull.

gy is released—often in the form of light. When the emitted light falls in the visible spectrum, we see it. When the stimulus is removed from the fluor, the light dims.

In the sea there are a great number of plants and animals with this characteristic. It was once thought that only inhabitants of deeper waters were fluorescent. But we have since learned that many plants and animals living in the shallows fluoresce.

The ability to fluoresce varies between species, and even between individuals of a species. In some corals, for example, only the limestone castle protecting the living polyp glows under the influence of our lamps, while in other corals just the polyp glows.

Chapter VIII. Eyes

Eyes usually come in multiples of two, except in those cyclopean creatures with only one eye. Some flatworms have hundreds of clusters of two- or three-celled "eyes." Light is perceived by these animals, but vision is not possible.

In lobsters, crabs, and many other crustaceans, as in many insects, the eyes are compound. That is, many tiny eyes make up a single large eye. Each eye has a lens and sends the light message back to the light sensitive tissue at the rear of the larger

> "Some flatworms have clusters of two- or three-celled 'eyes.' Light is perceived, but vision is not possible."

compound eye. There, like a mosaic, the various images are pieced together into a single image which the animal perceives.

As a rule, animals that live in the sea have spherical lenses that provide them with sharp vision underwater. Terrestrial animals, on the other hand, have elliptical lenses in their eyes. Among fishes a variety of eyes is found, each an adaptation to the special circumstances of that species. Anableps, the four-eyed fish, is not really four-eyed. It has only two eyes which are adapted to see above and below the surface of the water simultaneously as it lies at the surface. Some fishes that live their entire lives in lightless caves have no eyes, no way of perceiving light. Other cave-dwelling fish have vestigial eyes that may be able to sense light only dimly. Similarly, species of deep-sea fishes living in the darkness of the depths have tiny eyes that barely serve any function. Others of the deep oceans have large eyes with the capacity for opening

pupils to a great size, enabling them to gather what minute amounts of lights are available. Many deep-sea fishes can multiply the light their eyes receive and see even in the pitchblack areas of the deep sea. They do this by reseeing the same light—that is, light reflects within their eyes and they sense that same light several times.

Frogs, aquatic reptiles like the turtles and crocodiles, and the aquatic birds that dive for their food have a transparent third eyelid that slides across the eye to protect it underwater. The aquatic mammals are equipped with the necessary spherical lens.

> "Many deep-sea fishes can multiply the light their eyes receive and see even in the pitchblack areas of the deep sea. They do this by reseeing the same light—that is, light reflects within their eyes and they sense that same light several times."

Some of them, like the sea otter, that need vision in and out of the water have adjustable lenses that can change shape as needed. Human beings needing devices like face masks to see clearly underwater cannot yet be considered aquatic mammals!

Marine flatworm. The top photograph is of a flatworm, whose highly light-sensitive eyes are tiny two- or three-celled organs. Their eyes—and they may have hundreds of them along the sides of their bodies—have no lenses and they see no images; they only sense varying intensities of light.

Scallops. This animal has many eyes which rim the periphery of its mantle just inside the fluted edge of its beautiful shell, but they see little more than the flatworms. A few of the 40 or more of the tiny blue eyes of the bay scallop are shown here.

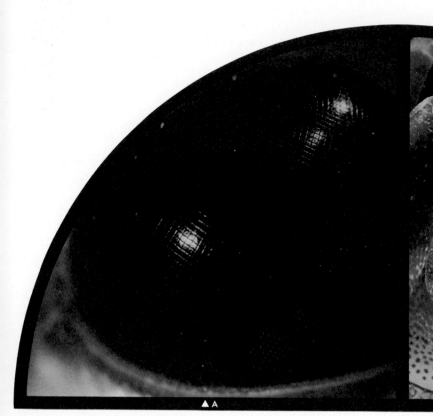 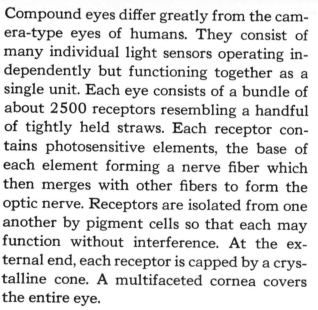

▲ A   ▲ B

The Compound System

A lobster's eyes are at the tip of a movable stalk, able to scan the whole vicinity. In a close-up we see a grid of fine lines dividing the animal's compound eyes into a large number of separate facets. These are ancient eyes—the first examples of which are seen in fossil rocks dating back 500 million years to the Cambrian period. The now-extinct relatives of the crustaceans, the trilobites, had them. Today many crustaceans and insects have them, although some of those forms are eyeless or bear different kinds of multiple eyes. To show us what we think a lobster may see, the photographer removed the lens from a lobster's eye. He placed this in front of the lens of his camera and took the picture of a sea star. The blurred portion was through the lobster's lens. The portion in sharp focus is as we see it.

Compound eyes differ greatly from the camera-type eyes of humans. They consist of many individual light sensors operating independently but functioning together as a single unit. Each eye consists of a bundle of about 2500 receptors resembling a handful of tightly held straws. Each receptor contains photosensitive elements, the base of each element forming a nerve fiber which then merges with other fibers to form the optic nerve. Receptors are isolated from one another by pigment cells so that each may function without interference. At the external end, each receptor is capped by a crystalline cone. A multifaceted cornea covers the entire eye.

Each of the many facets is a lens for one receptor. The unit consisting of one receptor, its crystalline cone, and the single facet of the corneal lens is called an ommatidium.

112

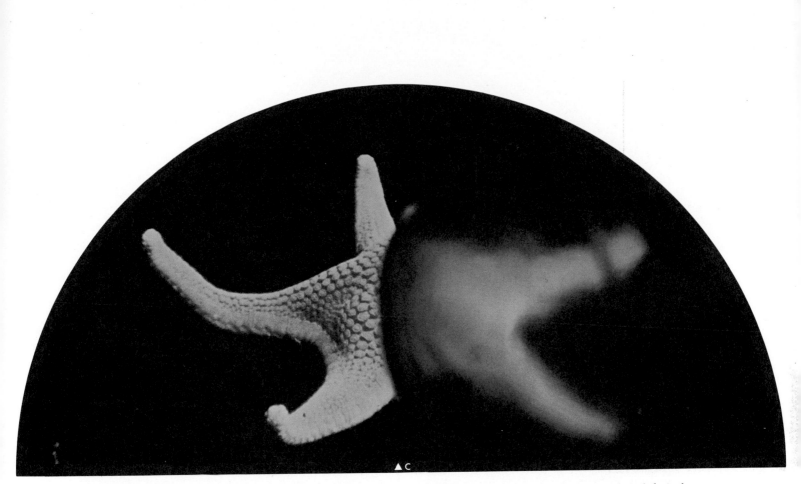

▲ C

Sleeves of pigments surrounding each ommatidium prevent the escape of light. In poor lighting these sleeves retract, permitting maximum use of available light, since light passes freely from one ommatidium to another until reaching a nerve fiber. Light entering any facet merges with light entering other facets. The information then is somewhat disorganized, since what is "seen" by one ommatidium may be transmitted to the brain by another's nerve fiber. Vision is less precise, but poor vision is better than none. In deep-sea crustaceans, living in the virtually lightless aphotic zone, the eyes are heavily endowed with reflective pigments which enable them to utilize the tiny amounts of light more efficiently.

Compound eyes are well adapted for detecting movement. Each ommatidium points in a slightly different direction than those

A/Close-up. In this close-up of a lobster's compound eye we can see the netlike grid which divides the eye into its many facets.

B / Head of a lobster. In this photograph of a lobster's head we can see the animal's eyes at the end of a movable stalk.

C/Two views of a sea star. In an experiment aimed at finding out what a lobster sees, the photographer removed the lens from the eye of a lobster and placed it in front of part of his camera lens.

around it and may perceive an event that by itself could be meaningless. But when the brain integrates hundreds of signals, the sum of the images is believed to produce a mosaiclike image. The lobster brain's ability to interpret the information has not been determined, but it is probable that crustaceans see more clearly than photographs like this one through its lens indicate.

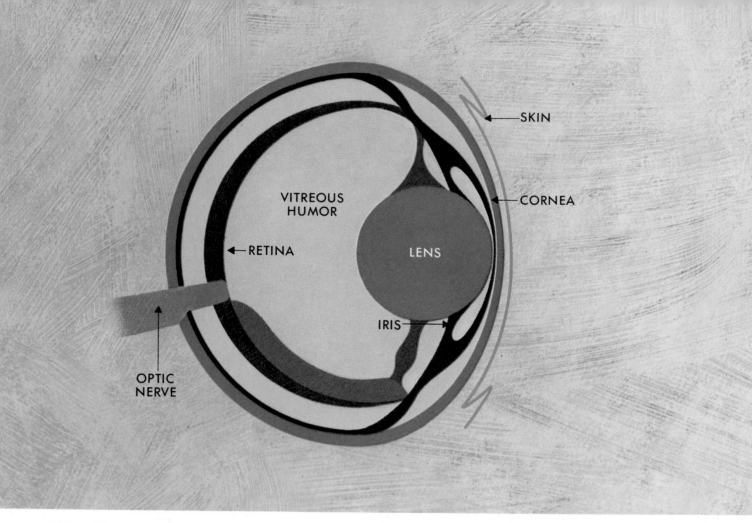

Labels on diagram: SKIN, CORNEA, VITREOUS HUMOR, RETINA, LENS, IRIS, OPTIC NERVE

The Eye of the Fish

A fish's eye view? We have no way of knowing exactly what a fish perceives. But it could be images similar to these pictures photographed through the lens of a red snapper's eye. The jellyfish is a pelagia and the fish is one of the sea basses, both from the Pacific Ocean. Although it may seem, because of the curvature of their lenses, that fish are nearsighted, experiments show that in fact they are more farsighted when seeing laterally through their lenses than when peering forward. They appear to have sharper vision looking forward, when they need it most.

Since most fishes, including the red snapper, have no muscles to modify the shape of their eye lenses, they move the entire lens backward or forward and focus within limits this way. This is exactly what we do when we focus a camera. To understand how a fish really sees, we must have a good knowledge both of the anatomy of the eye and of its general behavior.

Man has a spherical eye; a fish has a flatter one. Man's lens is oval and can change its curvature for near vision. The fish lens is spherical and hard. It does not change its shape but changes its position in the eye, moving forward to view near objects and backward to see things at a distance. The jellylike vitreous humor of man's eye is firm and supports the lens; the fish's vitreous humor is liquid.

A/Anatomy of a fish eye. The fish eye has a spherical lens that bends the light entering through it sufficiently for the image to come into focus on the retina.

B, C/Through the fish's eye. As seen through the lens of the red snapper's eye, the jellyfish, above, and sea bass, below, are perfectly visible.

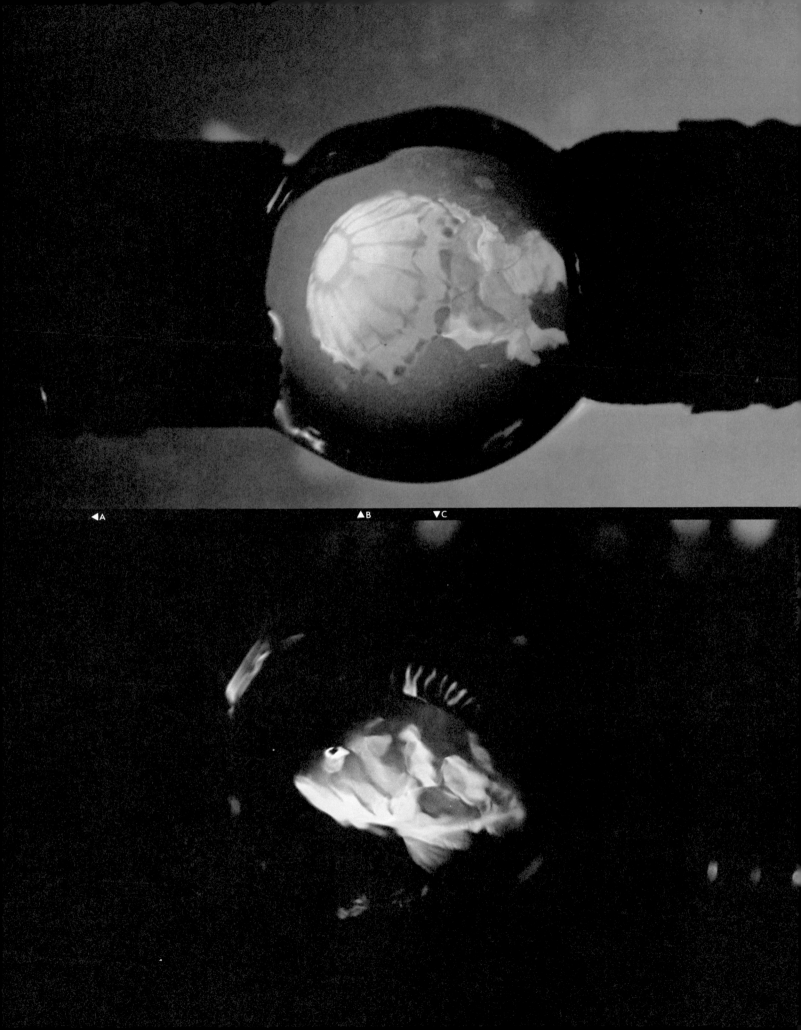

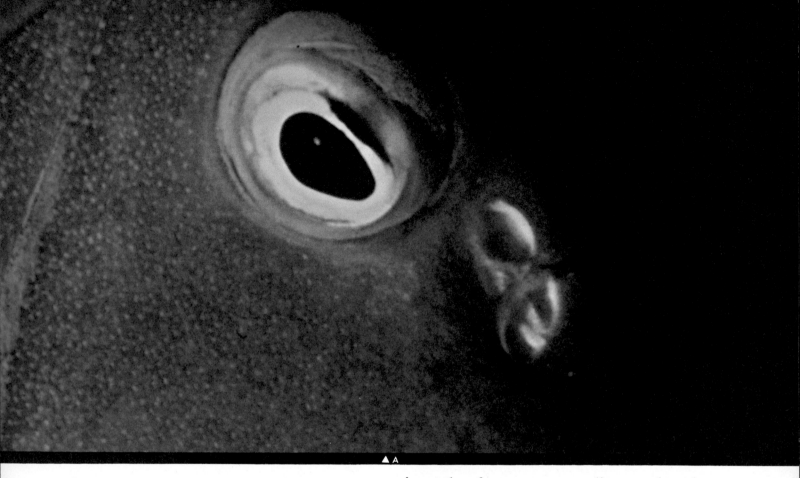

▲ A

Directing Vision

Like a Picasso painting, the flounder has two eyes on one side of its head—the side facing toward the surface. When the fish flopped over to live its life on one side, the eye on the under side migrated so it could function. The queen angelfish, on the other hand, has an eye on either side of its head, giving it two separate images. The ability to make its lenses bulge out and swivel each eye forward or backward gives this fish virtually 360° vision. Although the whale's vision would be more effective if the whale could see directly ahead and thereby have stereoscopic vision for better distance-judg-

ing, other factors seem to dictate a less desirable eye position. Many whales have relatively small, recessed eyes on the sides of their heads.

A/Queen angelfish. This fish has an eye on each side of its head, and it therefore sees two images at the same time.

B / Whale. Its eye is small in proportion to the size of its body. In many whales the forward vision is severely limited due to the positioning of the eyes.

C/Flounder. This fish, which lives flat on the ocean bottom, has both eyes on the uppermost side.

D / Blenny. This flatfish, which spends most of its life in the ocean, has its eyes placed close to the front and center of its head, allowing it excellent forward vision.

▲ B

▲ C ▼ D

▲ A

Positions of Eyes

The cephalopods, those shell-less molluscs that include the squid, octopus, and cuttlefish, have well-developed camera eyes located high and far forward. This eye position serves the dual purpose of giving them stereoscopic binocular vision forward and above. Coupled with its ability to change color and move quickly, the cuttlefish's eyes are a major defense system. In the pinkish sea robin projections surrounding the eye protect it.

A/Cuttlefish. The eye of this animal is positioned to allow it to see both forward and above.

B/Red hermit crab. This animal's eyes stick up away from its body, which is often hidden or buried in the mud, allowing it to see without being seen.

C/Armored sea robin. This fish can see two separate images—one on each side of its head.

▲ B   ▼ C

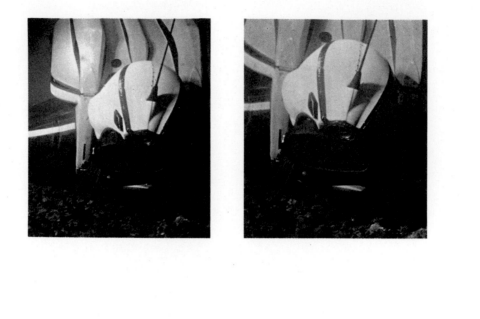

Binocular-Monocular Vision

As most fish swim along, they move their neckless bodies from side to side, scanning as they go. Fish use their two eyes independently for such scanning, which gives them monocular vision, each eye encompassing a wide range of 150°. But fish can face an object directly by swiveling their eyes forward, aimed almost in the same direction. This is binocular vision.

Binocular vision. The diver shown above is seen in front of the fish with the fish's right and left eyes together. In order to get this effect, the fish must swivel its eyes forward. In normal position the fish would be focusing on the scenery to its right and left.

Arcs of vision. Because most fish have eyes that are set wide apart on their heads, they are said to have "monocular" vision. That means that each eye sees something different—the right eye sees things on the right side, the left eye sees things on the left —shown in this chart by the color green. Fish, however, have the ability to swivel their eyes forward in order to focus on something in front of them.

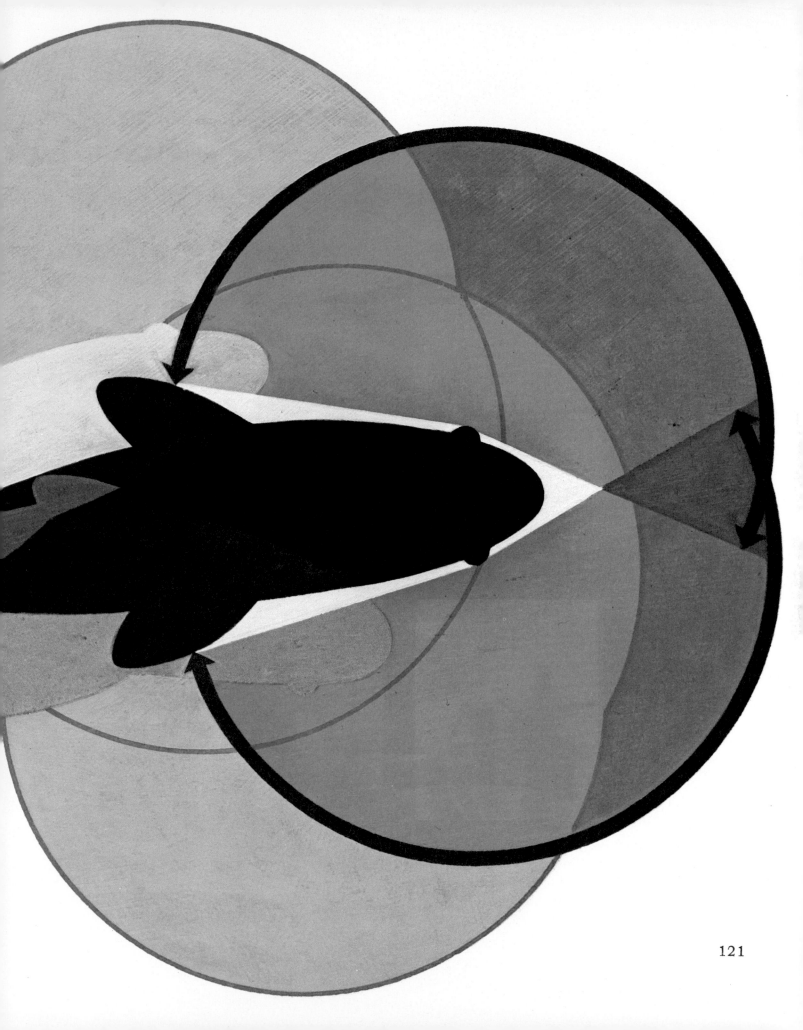

Shark's eye. *Above, a close-up of the eye of a shark. Behind the shark's retina is a group of mirrors reflecting any light entering the eye. The mirror bounces the light back to the retina. Thus the amount of light has the effect of being doubled.*

Deep-sea fishes. *In the drawing at right we see the* Gigantura, *which can see forward; the larval indiacanthus, whose eyes, at the ends of stalks, allow it to see virtually 360 degrees; and the* Opisthoproctus, *whose eyes are positioned to enable it to see upward.*

Vision in Darkness

Most of us are familiar with the glow of a cat's eyes caught in the beam of a car's headlights. Many fish also have reflecting eyes. Sharks have particularly spectacular ones. These eyes have an iridescent layer called the tapetum lucidum (bright carpet), a series of precisely placed mirrors behind the retina of the animal's eye. They enable the animal to see in very dim light. If a weak ray of light passes through a shark's retina, it may stimulate only one visual cell. The mirror behind the rod cell picks up the light and sends it back to the retina at such an angle that it stimulates another rod in the same cell grouping.

Abyssal residents have adapted in many ways to the rigors of life in the darkness of the deep oceans. Their coloration differs from near-surface dwellers. Many are lumi-

nescent, bringing their own light (and the only light) to the deeps. Finally, their eyes have had to evolve in unique ways. Most abyssal residents have large eyes, orbits occupying more than half the length of the animal's skull. But the size of the eye means little unless the pupil is enlarged too, allowing more light to enter the eye.

Then, of course, large eyes are often associated with large retinas and with many more receptor cells in the retina, which in turn give greater visual acuity. An eye is like a camera; while the size of the camera may influence the quality of the pictures, it has no bearing on the luminosity of the lens in dim light. The critical factor is the area of open iris. So, to improve vision, the pupils of deep dwellers must be proportionately larger than those of shallow dwellers. This permits more light to enter the eye.

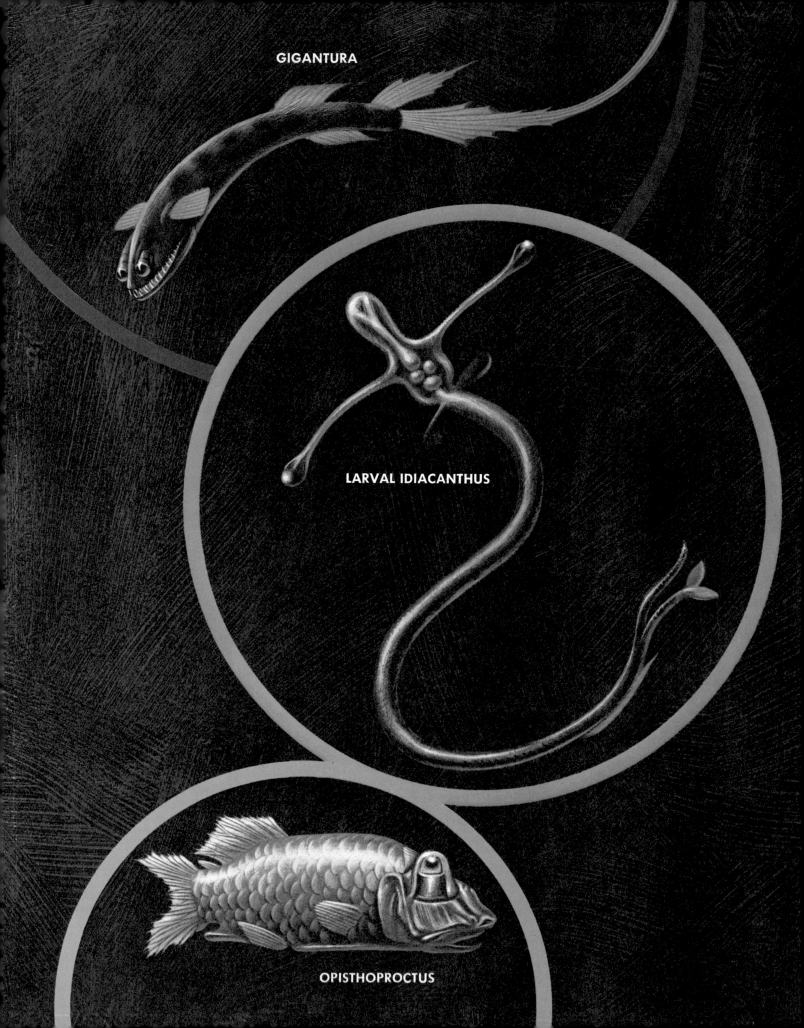

GIGANTURA

LARVAL IDIACANTHUS

OPISTHOPROCTUS

Horned puffin. The horned puffin, above, has a transparent eyelid which closes when it enters the water and acts as a correcting lens.

Cormorant. This bird has eye muscles so powerful that it can change the shape of the lens for undersea viewing, then change it back for surface sight.

The Bird's Eye

Fish-eating birds that dive for their dinner need clear underwater vision. This horned puffin has an adaptation for undersea viewing. A transparent eyelid, called a nicitating membrane, closes over the eye when it dives and bends the light sufficiently so it can see. With others it serves as a protection against dirt and dust. The cormorant has still another adaptation for underwater sight, powerful eye muscles that squeeze to change the shape of the lens toward the spherical for underwater and the elliptical for surface sight. The cormorant eye has probably the greatest range of focusing power of any animal—two or three times that of a human infant.

124

Sea Otter's Eyes

Vision is a primary hunting tool for the sea otter as he seeks sea urchins, crabs, and abalone in the dim light of the ocean floor 100 feet or more down. And despite his excellent underwater vision, he has good sight in the air too—an unusual combination. The reason he sees well in both air and water is that, like the cormorant shown on the previous page, he has a powerful muscle in each iris. This muscle can squeeze the lens and change its shape from spherical for underwater vision to elliptical for seeing in air. Turtles and some diving seabirds can also accommodate for equally good vision in both media—something the human species is not adapted to do.

Weddell Seal's Eyes

What visual adaptations enable amphibious creatures like seals and sea lions to see clearly in or out of water? Just as fishes have spherical lenses for underwater vision, so do seals and sea lions. To see in the air, most pinnipeds appear to be able voluntarily to alter some of the optical characteristics of their eyes. But scientists today are still divided as how exactly these alterations are achieved. Eyes of seals and sea lions are generally much larger than man's. In the Weddell seal the eyeball is almost double the dimensions of a human eye. And in whales, dolphins, and porpoises the eyes are still larger. Can sight mean more to them than to man? Perhaps.

Chapter IX. Fooling the Eye

Animals may live longer if they can fool the eyes of their enemies.

For many fish, camouflage means virtual invisibility—to prey and predator alike. It may mean disguising yourself to look like something you aren't. Or countershading, or obliterative shading. The barracuda and tarpon are darker above and lighter colored below. As one of these fishes cruises the open waters, sunlight from above brightens the darker back while the darkness of the depths tends to darken the brighter belly. Result: the fish is barely discernible, blending into the background of blue. In cryptic coloration, the creature blends with its background by being of the same color and pattern: flounder on a gravel bottom or trumpetfish among branches of staghorn coral. Some sea urchins cover their spines with bits of debris from their immediate surroundings; the decorator crab carries a menagerie on his carapace, which hides him from view.

Directive marks are an unusual defense against attack. Some animals carry a mark that directs attention away from them or away from their more vulnerable parts. Fake eyespots near the tail of some fish are assumed to draw attention from the head. In disruptive coloration, the readily recognizable outline of a fish is so broken up that a predator may not recognize the creature for what he is. The Moorish idol of the Pacific reefs and the jackknifefish of the Caribbean have bold, broad stripes which serve such a function. The icefish of the antarctic survives by remaining almost invisible because of its transparency. Even its blood is whitish; it carries no pigment.

In mimicry, an animal looks like another organism. Living among the weeds of the Sargasso Sea are many creatures that so resemble the yellowish brown algae that you have to look closely, even with the seaweed in your hand, or you will miss the shrimps, crabs, and sargassumfish hiding there. There's a wrasse that resembles the anemonefish in color and markings, if not in shape. It sticks close to anemones with their

> "Living among the weeds of the Sargasso Sea are many creatures that so resemble the yellowish brown algae that you have to look closely, even with the seaweed in your hand, or you'll miss the shrimps, crabs, and sargassumfish hiding there."

stinging tentacles, but not too close because it cannot mimic the anemonefish's immunity to their sting. Then there are the quick-change artists that switch colors or patterns to assume cryptic coloration. A flounder can match a checkerboard pattern.

A few fishes *want* to be seen. They advertise themselves so all may see them. It may be they need to be recognized by others of their own species. Or they advertise as a warning that they are dangerous, perhaps venomous creatures like the lionfish.

*The coloration of the **four-eyed butterflyfish** may to some extent fool an enemy's eyes. Near its tail is a dark spot contrasting sharply with the rest of its body. This spot resembles an eye. The real eye is concealed in a dark stripe on the fish's cheek. As these butterflyfish hover near a coral reef, they may seem vulnerable. But through their maneuverability, their tendency to hide in crevices, and possibly this deceptive body coloration, these fish thrive on reefs. Does this spot really fool a predator into misdirecting its attack to the posterior, giving the butterflyfish a better chance of escape? Behavioral evidence is limited, but it does seem logical.*

Disruptive Coloration

A painter seeks to capture the attention of an observer and then lead his eye around his painting until it reaches the focal point of the work. In nature the body art of some animals does exactly the opposite. Randomly placed splotches of bright colors, or stripes wandering about the sides of a fish, break up the outline and general shape of the animal, deceiving an observer into seeing something that isn't there or not recognizing what is there. This is called disruptive coloration. The vivid blue of the flagtail surgeonfish seems to make the fish a conspicuous member of the reef community, but the

black line winding along its sides divides the fish visually into several separate areas. A predator cruising the reef probably doesn't see the fish as a fish, but instead only as a series of random shapes moving along. The bold stripes of the jackknifefish perform in a similar manner. To another fish they probably appear to be just a series of lines.

Jackknifefish. The black and white stripes of the jackknifefish, above left, make it seem, in all probability, like a series of undulating lines to a predator rather than a meal.

Flagtail surgeonfish. Despite its beautiful blue color this fish, above right, is well protected from predators. Oddly shaped black lines on the fish's body make it hard to distinguish as a fish.

Countershading

Even in crystal-clear waters sharks appear suddenly, without warning. A look around may give a diver no hint of a shark's presence, yet an instant later one may be curiously circling only a few yards away. How can these large animals approach unnoticed when a diver has carefully scanned the surrounding waters? The answer is that sometimes we cannot see them even though they are nearby. Their sleek, muscular bodies are countershaded, dark on top and light underneath. The dark backs of sharks swimming

beneath us match the dark waters below them. Those swimming above us present a view of their light-colored ventral side, closely matching the lighted surface waters. We can better distinguish countershaded animals swimming at the same level as we are, since we can see the contrast of the light

Blue shark. *This animal, along with several other sea creatures, is "countershaded." The result: the animal is difficult to spot.*

underside with the dark dorsal side. But even then it is not easy. Countershaded species living in the open sea usually have blue or gray dorsal sides.

▲▲ A

Camouflage

For many fish, looking like something they're not means the difference between life and death. It keeps them hidden from both their predators and prey. In addition to the ones shown here there are, among others, the goosefish, the scorpionfish, and the flatfishes —who look so much like the ocean bottom that they are practically invisible!

A / Trumpetfish. *These fish have long tubular bodies and elongated snouts. Weak swimmers, they hover motionless, almost invisible, among gorgonian corals. Small, transparent dorsal and anal fins steady them as they hide and wait. Their heads may be up or down, and their large eyes take in all going on around them. Sometimes they align themselves with* *larger fish, like the remora does with sharks. Generally his companion is a herbivore or large fish that is not feared by the prey the trumpetfish is after. So the trumpetfish can approach his dinner unnoticed —using the companion for deception.*

B / Emerald clingfish. *These fish have evolved to an extraordinary degree, with pectoral fins modified into a sucker. With it, they attach themselves to rocks or seaweed and remain there. This one closely resembles the weed on which it is resting both in color and shape, even to the light spots on its back.*

C / Trunkfish. *This small fish is encased in a box of fused scales. The armor is like that of a turtle, and the fish has sacrificed mobility for the protection of the plating. To increase its chances of survival, the dark spots on the fish's light sides blend in with the coral background.*

▲ B   ▼ C

Chapter X. The Bright Palette

Many eighteenth-century minds were outraged by Isaac Newton's proclamation that sunlight is a mixture of all colors. It was tantamount to saying white is black. The white light of the sun contains all the colors of the spectrum; the white pigment of this page reflects all the colors of the spectrum.

Our eyes tell us that without light there is no color. A painter knows that color is an attribute of pigment molecules which reflect, absorb, or transmit specific waves of light. The book we see as red is made of pigments that absorb all colors except red. A green leaf reflects green, absorbs other colors.

To humans, color is a stimulant. It arouses the emotions or pacifies the mind. People, plants, animals, and objects are attractive to us because their colors are appealing, even more than because of their shapes.

When artificial light was introduced into the sea, it revealed brilliant colors there, colors never perceived by any eye till then, even by a fish's eye. Why are they there? What is their purpose? Do they *have* a purpose? We think we understand the reason for colors in the shallow waters that sunlight penetrates, but why do we find elegant chromatics, pinks, purples, mauves, and yellows below that? Why is this palette splashed across a world without sun?

Feather-duster worm. *Below, a feather-duster worm waves colorful appendages which emanate from its bright-orange center.*

Sea anemone. *A delicate mauve anemone waves its deep-green tentacles in the current. This exotic animal would make a fitting bonnet for the Madwoman of Chaillot.*

The Whole Spectrum

All the colors of the spectrum show up in the left-hand photograph taken with flash. Gorgonians—in lacy patterns of red, orange,

and yellow—join other types of coral animals dressed in pink and purple. A bicolor damselfish swims past the scene. In the right photo, the *Condylactis* sea anemone fluoresces with delicate pastel shades of green and purple.

The Beauty of Color

A giant sea fan undulates gracefully in the surge of the sea, a delicate tracery against the rich blue of the tropical ocean. Near its

base, surrounded by a stony coral, swarms a school of blue and yellow mojarralike fish. On the right page, in 100 feet of depth, a diver admires the bright-red shades of gorgonians flooded with artificial light. Without our lamps, the gorgonians would be blue.

The Gift of the Sun

From a planktonic sky
Starlight, chilled by a long journey
Mirrors on unruffled expanses around me.

        I don my dolphin skin.
        With humming head—bubble-elated,
        I slip into the cold flesh of the sea.

In anguish begins
The slow Fall of black angels
Brandishing blue suns.
In our beams are borne unlisted creatures
And colors intended to stay blind.

        Our chute ends on the dunes of Night
        Where I gather shreds of consciousness.
        Darkness is gravity
        My sky of joy is lost, above,
        Beyond 25 man-heights of liquid stone.

Ten eyes seek a glance, meet and reassure.
Suddenly heedful shadows leap
One torch lights on, throbs, floods with flickering gleams
A gaping chasm that has never known day.
Further is borne a second flattering star,
Then three, then six.
As our infernal squadron fins its gliding path
The everlasting night expires,
Pierced by the javelins we stole from the sun.

        The coral walls crumble, revealing
        Cloisters of frozen rain engulfed to oblivion.

Tomorrow the shine from the moon
Will frighten fish snouts into their burrows.
Tomorrow rain and hail will sprinkle the sea
With dancing diamonds
Tomorrow the vibrant sun will spread its light on the ocean.
A light to feed upon, to breathe—to fear—
The sun will stab the sea with flamboyant rays
And shimmer in the shallows,
Unable to cast a shadow.

        But my dreams will remain in the night
        In the deep Nautilus night.

Index

CHARTS AND ILLUSTRATIONS:

Howard Koslow—23, 26, 32, 33, 70, 71, 98, 106, 107, 114, 120, 121, 123.

PHOTO CREDITS:

John Boland—130, 134; Ben Cropp—129; David Doubilet—31, 94–95, 100–101; Jack Drafahl, Brooks Institute—28, 29, 52, 53, 64–65, 74, 92–93, 112, 113, 115, 117 (bottom), 122; Ruth Dugan—47, 99; Harold E. Edgerton—25, 78–79; Free-lance Photographers Guild: Howard Critchell—11, Gordon DeLisle—97, John Dowd—21, James Dutcher—12–13, Bob Gladden—77, 102, 109, L. Grigg—131, Dennis Hallinan—22, 57, Jerry Jones—18–19, Franz Lazi—127, Burton McNeely—27, C. G. Maxwell—75, Dennis Opresko—88, 89, 119 (bottom), John Stormont—48, 119, 135; George Green—85; Dr. C. Scott Johnson, Naval Undersea Center—132–133; Tom McHugh, Shedd Aquarium, Chicago, Ill.—117 (middle); The Mariners Museum, Newport News, Va.—30; Chuck Nicklin—116; NOAA Photo by Dr. Robert F. Dill—40, 41, 51; Photography Unlimited: Ron Church—87, 90–91; Photo Researchers: Russ Kinne—68–69; Carl Roessler—84; T. R. W. Systems: Dr. C. Knox and Dr. R. E. Brooks—104–105 (top); U.S. Navy, Naval Station, Washington, D.C.—44–45; R. Vahan—135 (bottom); Wards Natural Science Establishment, Rochester, N.Y., and Monterey, Calif.—111 (top); Ed Zimbelman—108.